RELIABILITY, STRESS ANALYSIS, AND FAILURE PREVENTION ISSUES IN EMERGING TECHNOLOGIES AND MATERIALS

presented at
THE 1995 ASME INTERNATIONAL
MECHANICAL ENGINEERING CONGRESS AND EXPOSITION
NOVEMBER 12–17, 1995
SAN FRANCISCO, CALIFORNIA

sponsored by
THE DESIGN ENGINEERING DIVISION, ASME

edited by
EROL SANCAKTAR
CLARKSON UNIVERSITY

THE AMERICAN SOCIETY OF MECHANICAL ENGINEERS
UNITED ENGINEERING CENTER / 345 EAST 47TH STREET / NEW YORK, N.Y. 10017

ISBN No. 0-7918-1763-6

Library of Congress Catalog Number 95-81068

FOREWORD

Polymer materials, polymer-based composites, adhesives, and piezoelectric materials continue to be among the emerging materials of importance in volume of industrial applications as well as materials of choice in critical applications in aerospace, aeronautical, and electronics, industries. For example, electronically conductive adhesives present an excellent alternative to the traditional lead soldering methods in electronic components, thus reducing the environmental concerns associated with the use of lead solders as well as providing directional conduction, rework capabilities, capability to apply in film form, etc.

Especially when these materials are in composite form, they present nontrivial problems in constitutive behavior, design, and analysis. The states of stress they are subjected to often contain geometric discontinuities and possess singularities. Because of this reason, computer-based simulation, and solution techniques involving numerical analysis and Finite Element Analysis (FEA) have become a mainstay of engineering analysis involving such complex problems.

Especially for the FEA as well as various numerical analysis techniques, the development of powerful, accurate, and user-friendly codes and their easy access will undoubtedly diffuse their use in engineering fields from analysis to all aspects of design and manufacturing. Various examples of such applications are presented in this book.

The Reliability, Stress Analysis and Failure Prevention (RSAFP) Committee of the Design Engineering Division participated in 1995 International Mechanical Engineering Congress and Exposition with a focus on emerging technologies and materials and fifteen papers presented in this program are contained in this volume. The sessions of the RSAFP program were entitled:

- Issues in Design Methodology and Reliability; Chairman; Dr. Erol Sancaktar;
    Vice-Chairman: Dr. John F. Maguire
- Electronically Conductive Adhesives; Adhesive Joints; Chairman: Dr. Erol Sancaktar;
    Vice-Chairman: Dr. Toshiyuki Sawa
- Lifetime Prediction of Polymers and Composites; Chairman: Dr. John F. Maguire;
    Vice-Chairman: Dr. John T. Bendler
- Application of FEA for RSAFP; Chairman: Dr. Zhanjun Gao;
    Vice-Chairman: Dr. Erol Sancaktar

As the RSAFP program organizer, I would like to extend my thanks to the session chairs and vice-chairs, as well as all authors and reviewers of the RSAFP program.

Erol Sancaktar

CONTENTS

An Electric Conduction Model for Electronically Conductive Adhesives
Yong Wei and Erol Sancaktar 1

On the Issue of Electric Boundary Conditions in Piezoelectric
Defect Problems
Horacio Sosa and Naum Khutoryansky 11

Two-Dimensional Thermal Stress Analysis in Butt Adhesive Joints
Containing Hole Defects and Rigid Fillers in Adhesive Under Non-Uniform
Temperature Field
Yuichi Nakano, Masahide Katsuo, Masataka Kawawaki,
and Toshiyuki Sawa 19

Two-Dimensional Stress Analysis and Strength Evaluation of Band
Adhesive Butt Joints Subjected to Tensile Loads
Toshiyuki Sawa, Yuichi Nakano, Katsuhiro Temma,
and Hiroaki Uchida 25

Elastoplastic Finite Element Analysis and Strength Evaluation of Adhesive
Butt Joints of Dissimilar Hollow Shafts Subjected to External
Bending Moments
Toshiyuki Sawa, Mitsuhiro Aoki, and Osama Nishikawa 35

A Comparison of Adhesively Bonded Single Lap, Scarf and Butt Joints
Jeffrey S. Baylor and Erol Sancaktar 41

The Effect of Fiber Type on the Level of Stress Concentration Created in
Filleted Composite Rectangular Bars in Bending
Mustafa Gür, Aydin Turgut, and Erol Sancaktar 49

On the Role of the Chemical Potential in Predicting Long Life Performance
of Polymers and Polymeric Composites
John F. Maguire and Peggy L. Talley 57

A Cumulant Expansion Approach to Predict the Service Lifetime of
Polymers and Polymer Composites
Michael A. Miller and John F. Maguire 63

Maintaining a Consistent Configuration in a Constraint-Based Mechanical
Design System
A. Vincent Huffaker and Oded Z. Maimon 67

The Conceptual Design of Roto-Cooler Air Conditioning System
Yuan Mao Huang 79

Failure Pattern of Machining Centers (MC)
Jia Yazhou, Wang Molin, and Jia Zhixin 89

Reliability Analysis of Machining Centers (MC)
Jia Yazhou, Wang Guiqin, Ma Jian, Jia Zhixin 95

Investigation of Nonlinear Behavior of Vehicle Body Panels Under
Shipping Conditions
Houchun Xia and David Y. Xue 99

Simulation of the Drop Test of a Packaged Refrigerator
Seoggwan Kim, Wonjoon Cho, L. T. Kisielewicz, A. Petitjean,
and K. Matsumura 109

Author Index 117

AN ELECTRIC CONDUCTION MODEL FOR
ELECTRONICALLY CONDUCTIVE ADHESIVES

Yong Wei and Erol Sancaktar
Department of Mechanical and Aeronautical Engineering
Center for Advanced Materials Processing
Clarkson University
Potsdam, New York 13699-5729

ABSTRACT

The mechanism of interparticle conduction is modeled theoretically and assessed experimentally. In the first part of this paper, we predict and illustrate that the electrical resistance of powders is dependent on the external pressure applied on them. A pressure dependent model is developed to predict interparticle electric conduction by representing interparticle contact resistance as the sum of constriction resistance and tunnelling resistance. The theoretical model developed is confirmed experimentally using 9 micron diameter spherical nickel particles under pressure. The effect of film thickness on the conduction behavior of electrically conductive adhesives is also presented. For comparison purposes, an analytical relation is developed to predict three dimensional resistivity of particle filled conductive adhesives. This analysis reveals that the adhesive's resistivity depends on parameters, m, representing an average contact number and, S_i, representing the average length of conductive paths between the conductive particles incorporated in the adhesive matrix. Both the parameters m, and S_i are functions of the conductive particle volume fraction ϕ, and these functional relations are developed by computer simulation for the cases of spherical particles in three dimensional (3-D) and two-dimensional (2-D) contact. The latter (2-D) case represents thin film applications.

INTRODUCTION

Conductive adhesives are usually two-phase composites with adhesive base matrix and conductive metal filler. Several different conduction mechanisms have been proposed for conductive adhesives. Among them, the percolation model is the most widely used one. We note that all of the models proposed so far are based on the assumption of a completely developed conductive path where electrons can go through. The development of an efficient and durable conductive path, however, still poses a problem to be solved. Obviously, since conductive path (or chain) is formed by metal particles, these conductive particles must initially make intimate contact (a general contact which includes the physical contact and the tunneling) under the action of some force so that, ultimately, they transport electrons. Therefore, it is necessary to define a connection force for the metal particles, understand its nature and relate it to electric conductivity in the adhesive. For this purpose, a comparative model, based on homogeneous powder materials, is proposed for evaluating the correlation between this connection force and conductivity. Using the concepts developed initially with this model, we can then attempt to predict the nature of conductivity in conductive adhesives. This paper also presents an initial study on the film thickness dependent conductivity of isotropic conductive adhesives.

Model for Conductivity

This model is based on a simple device used to test conductivity for homogeneous powders. The device consists of a hollow cylinder with inner diameter, $D = 11.86$ mm. Two short metal bars are used to transfer the external force (q) to the metal powders and, as the electrical poles for conductivity measurements, as shown in Figure (1). Experiments reveal that in the absence of an external compressive force, i.e. $q = 0$, electric current is not conducted between the two metal poles. As we introduce a compressive force, a current path is established in the metal powders and conductivity increases with increasing compressive force, as shown in Figures (2) and (3). Obviously, intimate contact in metal fillers is the crucial factor for efficient and durable electric conduction.

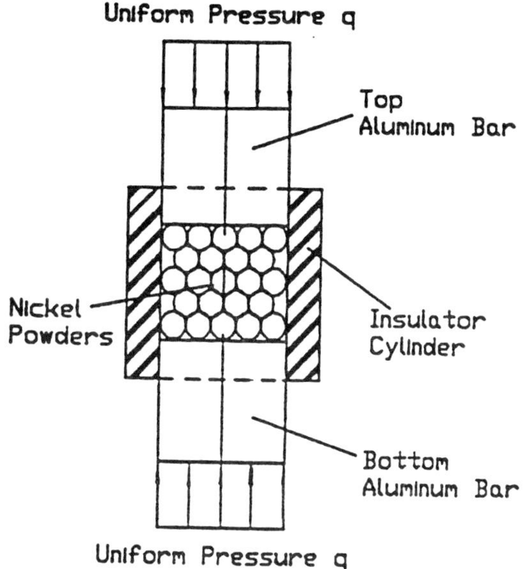

FIGURE 1: THE COMPARATIVE MODEL FOR CONDUCTIVITY

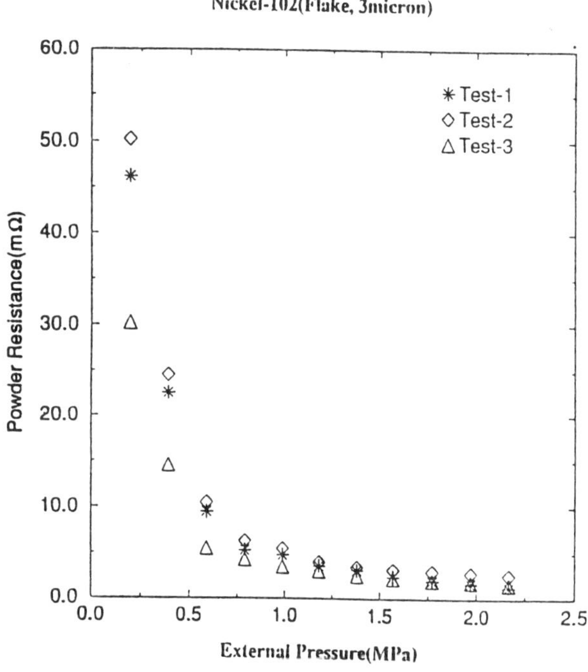

FIGURE 2: THE EFFECT OF EXTERNAL PRESSURE ON POWDER RESISTANCE FOR NICKEL-102 (RESISTANCE MEASURED ON 15MM LONG POWDER FILLED CYLINDER OF 11.9MM DIAMETER)

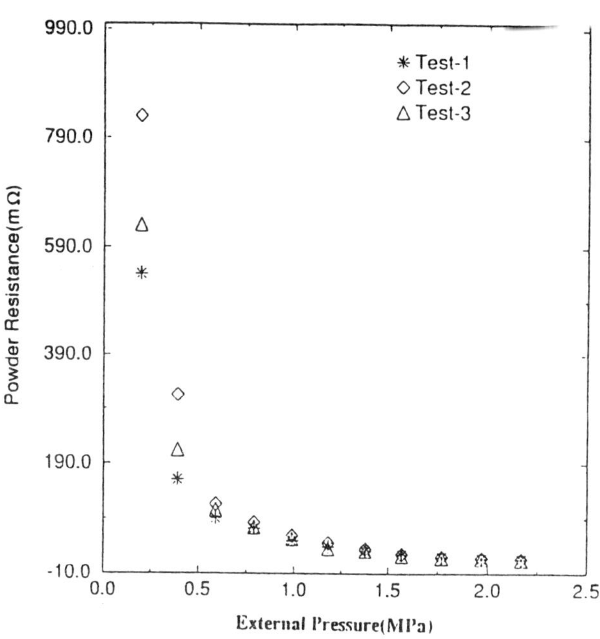

FIGURE 3: THE EFFECT OF EXTERNAL PRESSURE ON POWDER RESISTANCE FOR NICKEL-103 (RESISTANCE MEASURED ON 15MM LONG POWDER FILLED CYLINDER OF 11.9MM DIAMETER)

We assume that intimate metal to metal contact does not exist for 100% of this deformed area, A. Instead, only a part, say A_1, of the total area, A is in metal to metal contact, as shown in Figure (4) and, the remaining portion, A_2, contains intermediate layer between metal surfaces (Holm, 1967). For example, for

A-A Cross Section

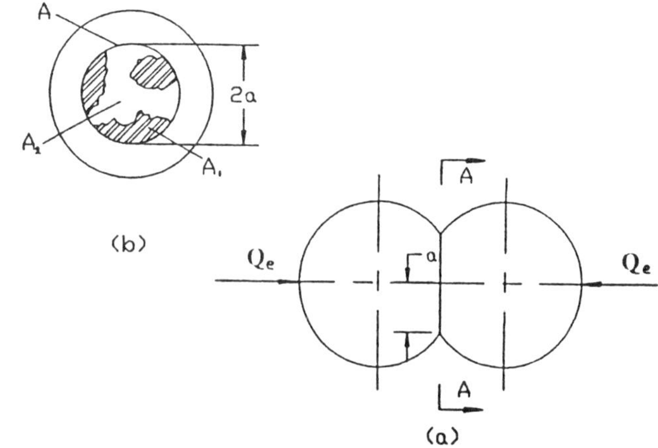

FIGURE 4: TWO SPHERES IN CONTACT UNDER DIAMETRIC PRESSURE (A) AND, THE RESULTING CONTACT AREA (B)

nickel powder in the compression device, this intermediate layer (s) consists of a mixture of nickel oxide and air. When the electrons go through this contact area, the resistance from area A_1 is called the constriction resistance R_c, and the resistance from area A_2 is called the tunnel resistance R_t. Of course, the intermediate layer between the metal spheres are considered to be thin films of several nanometer thickness or void volumes of comparable size. The total contact resistance R_s at this contact area A is the sum of the constriction resistance and tunnel resistance, i.e.

$$R_s = R_c + R_t. \qquad (1)$$

Interparticle Contact Force

In order to be able to establish a relation between the compression force applied on metal particles in contact and the resulting conductivity, we first need to relate the external pressure, q, of the compression device to the interparticle contact force. For this purpose we first make the following assumptions:

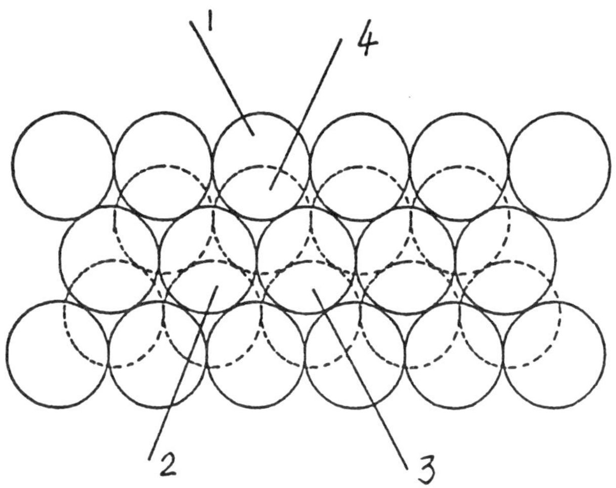

FIGURE 5: HEXAGONAL PACKING GEOMETRY (SPHERES WITH DOT LINES ARE IN THE TOP LAYERS)

(i) The metal particles in the compression cylinder are distributed with hexagonal packing arrangements.

(ii) The force is uniformly transferred from the top layer to the next layer in the axial direction.

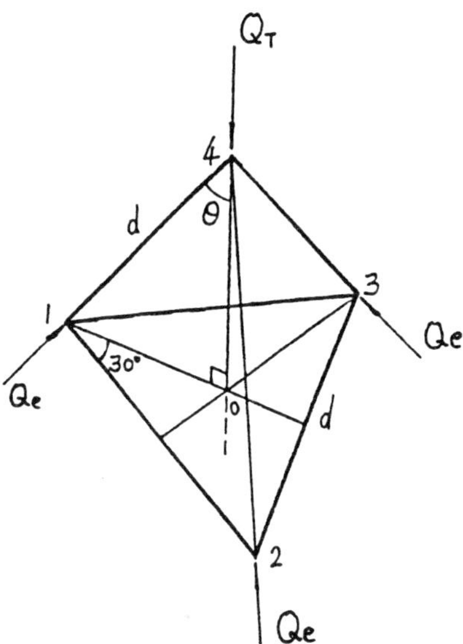

FIGURE 6: FREE BODY DIAGRAM FOR CALCULATING CONTACT FORCE

In our cylindrical compression device the total external pressure applied on the top layer is q (N/m^2). The uniform force, Q_T, applied on individual particles at the top layer is given by:

$$Q_T = q\pi r^2/f \qquad (2)$$

where r is radius of the particle and f is the packing factor, which is .906 in 2-D hexagonal packing.

Formation of the Contact Area

There are twelve contact points for each individual particle, except for the top and bottom layers in the cylinder. The contact force between particles is defined as Q_e. Let's look at four particles 1, 2, 3 and 4, where particle 4 is in the top layer and particles 1, 2, and 3 are in the layer next to the top one (Figure 5). Q_T will be balanced by Q_e (Figure 6). The angle between Y-axis and line 1-4 is $\theta = 35.3$.

$$(Q_e + Q_e + Q_e)\cos\theta = Q_T \qquad (3)$$

Therefore,

$$Q_e = 0.408Q_T. \qquad (4)$$

In the presence of contact forces Q_e, the metal spheres will deform and a contact area is formed. Using elasticity theory (Timoshenk and Goodier, 1970), the radius, a, of the circular contact area (Figure 4) is calculated as

$$a = \sqrt[3]{\frac{3Q_e(1 - v^2)r}{4E}} \qquad (5)$$

where υ is the Poisson's ratio, and E is the Young's modulus of particles.

By substituting Q_c in Equation (5), we get

$$a = \sqrt[3]{\frac{3 \times 0.408 \times q\pi r^2(1 - v^2)r}{4E}}$$

$$a = 1.02\sqrt[3]{\frac{qr^3(1 - v^2)}{E}} \qquad (6)$$

Electrical Resistance at the Contact Point

As mentioned before, the contact resistance is the sum of constriction resistance and tunnel resistance,

$$R_s = R_c + R_t$$

$$R_c = (\rho_1 + \rho_2)/4a, \qquad (7)$$

where ρ_1 and ρ_2 are the resistivities of spheres 1 and 2 respectively and,

$$R_t = \rho_\sigma / \pi a^2. \qquad (8)$$

In Equation (8), ρ_σ (Ω - m^2) defines the tunnel resistivity.

Substitution into Equation (1) yields

$$R_s = (\rho_1 + \rho_2)/4a + \rho_\sigma/\pi a^2. \qquad (9)$$

In our case, the particle resistivities are the same, i.e. $\rho_1 = \rho_2 = \rho$, consequently,

$$R_s = \rho/2a + \rho_\sigma/\pi a^2. \qquad (10)$$

We should note that the term ρ_σ of Equation (10) is a function of the tunnel width s, the work functions ϕ_1 and ϕ_2 for electron emission from sphere 1 and sphere 2 into the tunnel, the applied voltage V, and the relative permitivity ε_r of the material in the tunnel. For small tunnel width s ($< 15A$), ρ_σ is approximately independent of the voltage V (Holm, 1967). When the two sphere's materials are the same, i.e. $\omega_1 = \omega_2$, then

$$\rho_\sigma(s,\omega, \varepsilon_r) = (1/2) 10^{-22} (A^2/1 + AB) e^{AB} \; (\Omega - cm^2) \qquad (11)$$

with

$$A = 7.32 \times 10^5 (s - 7.2/\omega) \qquad (12)$$

and

$$B = 1.265 \times 10^{-6}(\phi - 10/s\varepsilon_r)^{1/2}. \qquad (13)$$

The Tunnel Width

As mentioned before, the tunnel resistance is a function of the tunnel width s. In our comparative model, the resistance doesn't significantly change when tunnel width s is less than 4A. But when s > 4A, the resistance will significantly increase with increasing tunnel width s, as shown in Figure 7. This Figure also reveals that the resistance will be less when the particle size is larger.

The tunnel width is basically the sum of the air gap in the interface between particles and the thickness of the nickel oxide, s_1, which is on the particle surface. With the application of external pressure, some air in the interface will be pumped out and the tunnel width, s, will be decreased. However, s can not be less than the metal oxide thickness s_1 on the particle surface, i.e. $s > s_1$. Consequently, a lower limit exists for tunnel resistance which when added to the resistance across metal particles, approaches the asympototic resistance level of the external pressure vs. powder resistance curve. Obviously, when the external pressure increases to the point where $s = s_1$, the tunnel resistance will reach a constant value, since the tunnel width is no more a function of the external pressure. Assuming that material contact exists over the tunnelling area and that linear contact constitutive behavior prevails, the relation between the tunnel width, s and the external pressure, q, can be considered linear, i.e.

$$s + C_1q + C_2 \qquad s_m > s > s_1 \qquad (14)$$

where s_s is a number not much bigger than s_1. For our work, we assume that $s_m = 13A$ so that a reasonable resistance value can be obtained in Equation (9). In Equation (14), C_1 and C_2 are experimentally determined constants which, in our case, are $C_1 = -2.31$ and $C_2 = 13.38$. Substitution of Equation (14) in Equation (11) results in an expression for the tunnel resistivity ρ_σ. The total contact resistance at one contact point can now be calculated as

$$R_s = 0.49\frac{\rho}{r} \sqrt[3]{\frac{E}{q(1 - v^2)}} + 0.31$$
$$\frac{\rho_\sigma}{r^2} \sqrt[3]{[\frac{E}{q(1 - v^2)}]^2} \qquad (15)$$

THE DEPENDENCE OF ELECTRIC CONDUCTION ON FILM THICKNESS

Simulation of Particle Contacts and Conductive Paths.

We consider a cuboid domain with three side lengths of $L_x \times L_y \times L_z$ (Fig. 8). We assume that there are k_i paths (i = x, y, z). k_i is defined as the number of conductive paths along the X axis. k_y and k_z are defined similarly. Suppose the conductive path length are S_i (i = x, y, z). Here we have S_i as the average length in i (i = x, y, z) direction.

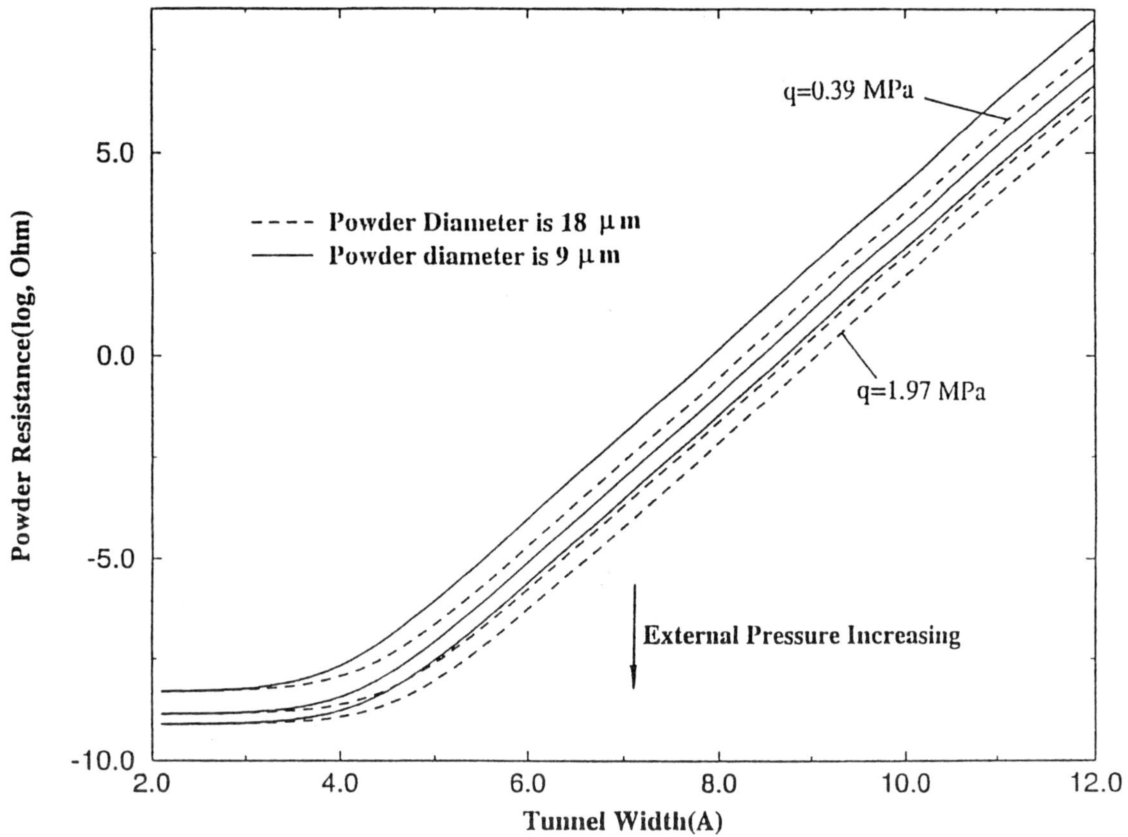

FIGURE 7: THE VARIATION OF POWDER RESISTANCE WITH TUNNEL WIDTH, PARTICLE DIAMETER AND, EXTERNAL PRESSURE (RESISTANCE CALCULATED ON 15 MM LONG POWDER FILLED CYLINDER OF 11.9 MM DIAMETER)

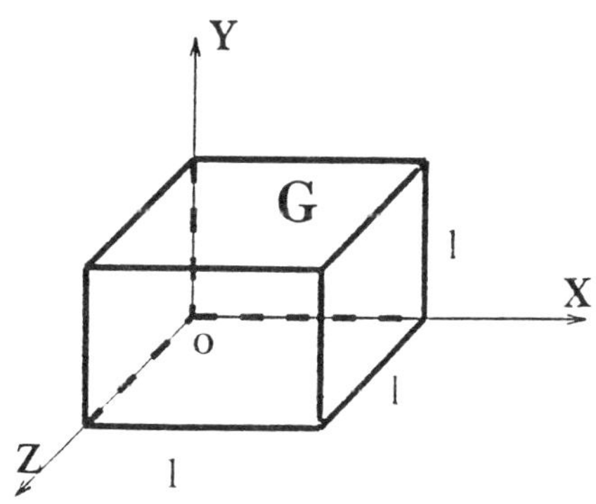

FIGURE 8 THE COMPUTER SIMULATION DOMAIN AND ITS GEOMETRY

By percolation theory, the conductive network can't be formed until the volume fraction of particle fillers reach the critical point, which is experimentally determined as 0.35 for spherical fillers (Bhattacharya, 1986). The average particle contact number ($\overline{m}$) at critical point (critical volume fraction) is about 1.5 (Bahattacharya, 1986).

With the volume fraction, ϕ, given, we can calculate the total particle number (N_T) in the chosen cuboid domain (G),

$$N_T = \frac{6L_x L_y L_z}{\pi D^3}\phi \qquad (16)$$

As one of the conductive path members, the particle in a conductive path has to have at least two contact points with its neighboring particles. This means that only $\overline{m}(N_T/2)$ particles will participate in conductive paths.

Suppose ρ_x, ρ_y and ρ_z are resistivities along X,Y,Z axes. Let's consider X-axis first. By definition of resistivity, we have

$$\rho_x = R_x \frac{A_x}{L_x} = R_x \frac{L_y L_z}{L_x} \qquad (17)$$

where

$$R_x = \frac{1}{k_x} r_x \qquad (18)$$

is the total resistance in the X direction and r_x is the resistance of each path in X direction which is assumed to be the same for each path. Let's assume that contact resistance between two particles is R_s which includes tunnel and constriction resistances as given by Equation (15). Then,

$$r_x = (\frac{S_x}{D} - 1)R_s, \qquad (19)$$

where, S_x, is the average length of conductive paths. Since $(S_x/D) \gg 1$, $(S_x/D - 1) = S_x/D$. Then,

$$R_x = \frac{1}{k_x} \frac{S_x}{D} R_s. \qquad (20)$$

Therefore,

$$\rho_x = \frac{1}{k_x} \frac{S_x}{D} R_s \frac{L_y L_z}{L_x} \qquad (21)$$

similarly,

$$\rho_y = \frac{1}{k_y} \frac{S_y}{D} R_s \frac{L_x L_z}{L_y} \qquad (22)$$

and,

$$\rho_z = \frac{1}{k_z} \frac{S_z}{D} R_s \frac{L_x L_y}{L_z}. \qquad (23)$$

The average length of conductive paths can be defined as $S_x = CL_x$, where C is a conductive path factor, which is a function of volume fraction, i.e.:

$$C = f_x(\phi). \qquad (24)$$

Statistically, the conductive path factor is inversely proportional to volume fraction ϕ. This implies that the average conductive path's length increases with decreasing volume fraction, as shown in Figure (9). Note that, similarly we have $S_y = CL_y$, and $S_z = CL_z$.

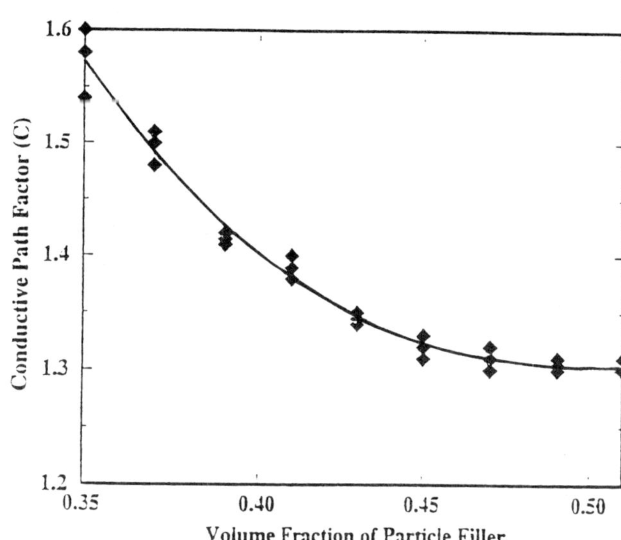

FIGURE 9 THE RELATIONSHIP BETWEEN CONDUCTIVE PATH FACTOR C AND VOLUME FRACTION Φ

Hence,

$$\rho_x = \frac{1}{k_x} \frac{CL_x}{D} R_s \frac{L_y L_z}{L_x} \qquad (25)$$

i.e.,

$$\rho_x = \frac{C}{k_x} \frac{R_s}{D} L_y L_z \qquad (26)$$

Similarly,

$$\rho_y = \frac{C}{k_y} \frac{R_s}{D} L_x L_z \qquad (27)$$

and,

$$\rho_z = \frac{C}{k_z} \frac{R_s}{D} L_y L_x. \qquad (28)$$

In Equations (26, 27, 28) there are four unknowns k_x, k_y, k_z and ρ if we have isotropic properties, i.e.,

$$\rho_x = \rho_y = \rho_z = \rho. \qquad (29)$$

Hence, we need an additional equation to solve them. All particles in conductive paths should have two contact numbers for each. Therefore, the total contact numbers of all particles in conductive paths should be equal to $\overline{m}(N_T/2)$. Therefore, we have:

$$k_x(S_x/D) + k_y(S_y/D) + k_z(S_z/D) = \overline{m}\frac{N_t}{2}. \quad (30)$$

Using Equations (29,30), we can solve for k_i (i = x, y, z).

$$\rho_x = \rho_y \;\rightarrow\; \frac{L_y L_z}{k_x} = \frac{L_x L_z}{k_y} \;\rightarrow\; k_y = \frac{L_x}{L_y}k_x \quad (31)$$

and,

$$\rho_x = \rho_z \;\rightarrow\; \frac{L_y L_z}{k_x} = \frac{L_x L_y}{k_z} \;\rightarrow\; k_z = \frac{L_x}{L_z}k_x. \quad (32)$$

Inserting k_y and k_z into Equation (30), we have

$$k_x(S_x/D) + \frac{L_x}{L_y}k_x(S_y/D) + \frac{L_x}{L_z}k_x(S_z/D) = \overline{m}\frac{N_T}{2}. \quad (33)$$

Using $S_x = L_x \times C, S_y = L_y \times C$, and $S_z = L_z \times C$, in Equation (33) we get:

$$3k_x C L_x/D = \overline{m}\frac{N_T}{2} \quad (34)$$

and,

$$k_x = \frac{\overline{m}N_t D}{6CL_x} = \overline{m}\frac{L_y L_z \phi}{\pi D^2 C}. \quad (35)$$

Similarly,

$$k_y = \overline{m}\frac{L_x L_z \phi}{\pi D^2 C}, \quad (36)$$

and,

$$k_z = \overline{m}\frac{L_x L_y \phi}{\pi D^2 C}. \quad (37)$$

Therefore,

$$\rho = \rho_x = \frac{C}{\overline{m}\dfrac{L_y L_z \phi}{\pi D^2 C}}\;\frac{R_s}{D}\,L_y L_z \quad (38)$$

i.e.,

$$\rho = \frac{C^2 \pi D R_s}{\overline{m}\phi} \quad (39)$$

where D is the diameter of spherical filler, R_s is contact resistance (Equation 15), ϕ is volume fraction of fillers, and m is the average contact number. In order to obtain an estimate of the resistivity using Equation (39) we first need to obtain an expression for the terms $\overline{m}$ and ϕ. This can be done by using a computer simulation program details of which are given in reference (Wei, 1995). Using this computer simulation we get:

$$\overline{m} = 0.1625 + 5.7323\phi - 5.1716\phi^2 \quad (1.5 < \overline{m} < 2.0)$$
$$(40)$$

which is graphed in Figure (10).

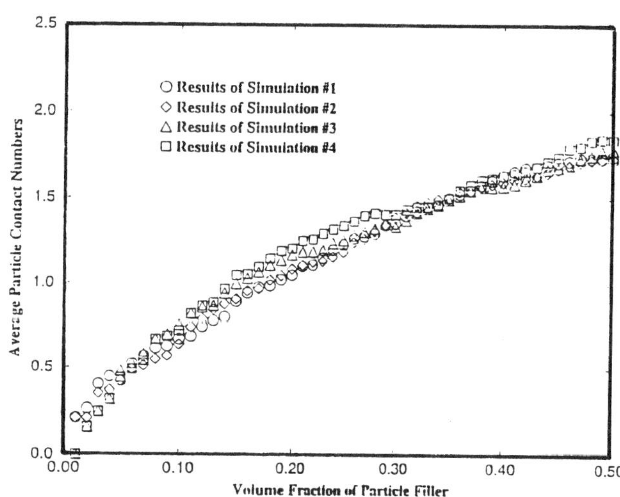

FIGURE 10 SIMULATION RESULTS FOR AVERAGE CONTACT NUMBER M

Conduction Simulation of Thin Conductive Adhesive Film

As discussed earlier, conductivity is established by the metal conductive paths through which the electrons move. These paths are developed in 3-D, referred to as 3-D paths. The average length (S) of the paths follow the statistical formula:

$$S = C \times L \tag{41}$$

where L is the side length of a unit cuboid along the direction considered and C is conductive path factor with values $1.3 < C < 1.6$. This conductive path factor is also a variable to the volume fraction, ϕ, of the particle fillers.

For thin adhesive films (e.g., $z < 150\mu m$), the development of conductive particle paths is approximately in two dimensions, instead of the three dimensions discussed above. The unit cuboid side length L_z considered infinite in 3-D becomes finite compared with the size of particle fillers. In this case, the conductive metal paths develop in 2-D planar direction rather than in 3-D. By using computer simulation we can show that the conductive path factor C' in a thin layer adhesive film differs from that in three dimensional simulation. As shown in Figure (11), at lower volume fraction of metal-filler, the difference between the C in three dimensions and the C' in two dimensions is very large. At higher particle volume fractions the difference between C and C' becomes smaller. This implies that with high volume fraction of fillers, the average length of conductive path is about the same for both three dimensional and two dimensional conductive adhesive applications.

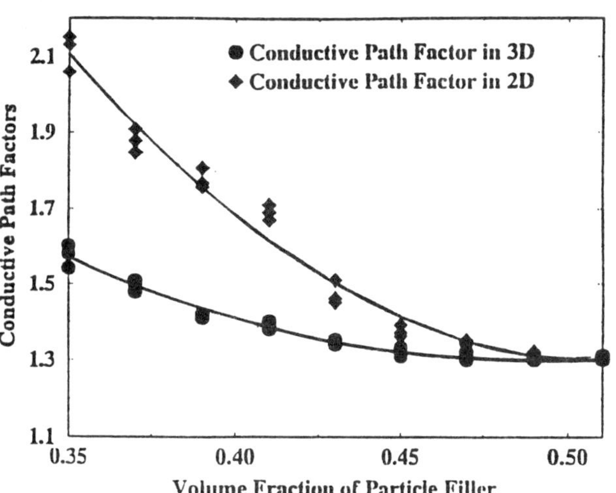

FIGURE 11   CONDUCTIVE PATH FACTORS C AND C′ AS OBTAINED BY COMPUTER SIMULATION

EXPERIMENTS

Four different commercial conductive adhesives with labels Adhesive A, Adhesive B, Adhesive C and Adhesive D, respectively, were used to make experimental samples. The properties of these adhesives are shown in Table (1). All these adhesives contain silver particles which are essentially in flake configuration. Experimental samples were prepared by using cure schedule and procedures suggested by the manufacturers. Eight different thicknesses groups were prepared for each conductive adhesive. Three samples were prepared in order to obtain average values. Adhesive thicknesses were measured using DEK style profile meter. The film resistances along the longitudinal direction were measured using the four wires method and the results are shown in Figures (12) through (15).

SUMMARY AND CONCLUSIONS

Theoretical and experimental methodology was developed in order to understand the effect of external pressure on the conductivity of homogeneous metal particles. A hollow cylinder with conducting plungers was used to illustrate that the conductivity of metal particles increases with increasing compressive force. This behavior was described analytically based on the constriction and tunnel resistances. The resistance between particles is governed by Holm's equation, which considers total resistance in two parts: constriction resistance, and tunnelling resistance.

The analysis presented also reveals that particle filled electrically conductive adhesive's resistivity depends on a parameter, S_i , representing the average length of conductive paths between the conductive particles incorporated in the adhesive matrix. The computer simulation utilized reveals that the S_i values in the planar direction of thin film adhesives is much larger than that for the 3-D case. This leads to anisotropic electric conduction in thin film adhesives where the resistivity values in planar directions are much larger than those encountered in 3-D measurements for the same adhesive material. This behavior is illustrated using four different commercial conductive adhesives containing silver flakes.

ACKNOWLEDGEMENTS

Financial support by IBM Corporation and New York State Science and Technology Foundation through the Center for Advanced Materials Processing at Clarkson University is gratefully acknowledged. We also would like to acknowledge an ISHM Educational Foundation Research Grant to Y. Wei for 1994-1995.

REFERENCES

Holm, R., 1967, Electric Contact, Springer-Verlag Inc., New York.

Timoshenko, S.P., Goodier, J.N., 1970, Theory of Elasticity, McGraw-Hill Book Co., New York.

Bhattacharya, S.K., 1986, Metal-Filled Polymers-Properties and Applications.

Wei, Y., 1995, Electronically Conductive Adhesives: Conduction Mechanisms, Mechanical Behavior and Durability, Ph.D. Dissertation, Clarkson University, Potsdam, New York.

TABLE 1. PROPERTIES OF MODEL ADHESIVES

Samples	Conductivity (Ω - cm)	Viscosity @25C(cps)	Cure Temperature (°C)
Adhesive A	6×10^{-5}	55,000	130 - 150
Adhesive B	$< 5 \times 10^{-4}$	200,000	80
Adhesive C	1×10^{-4}	220,000	130 - 150
Adhesive D	1×10^{-4}	200,000	170 - 210

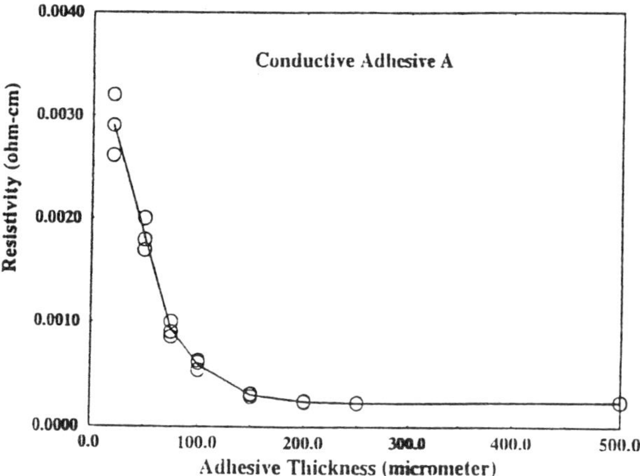

FIG. 12 RELATIONSHIP BETWEEN RESISTIVITY AND ADHESIVE FILM THICKNESS FOR CONDUCTIVE ADHESIVE A

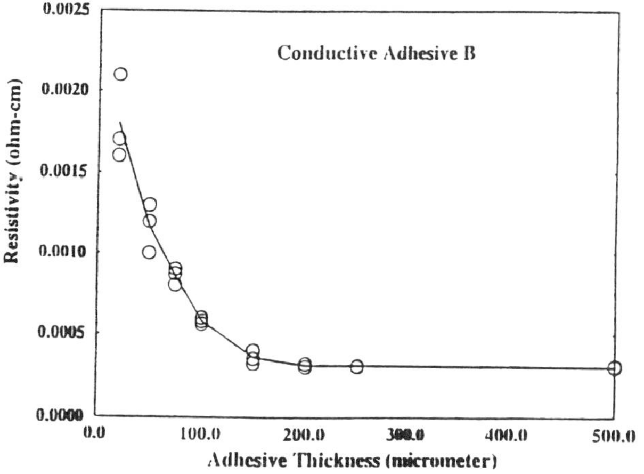

FIG 13 RELATIONSHIP BETWEEN RESISTIVITY AND ADHESIVE FILM THICKNESS FOR CONDUCTIVE ADHESIVE B

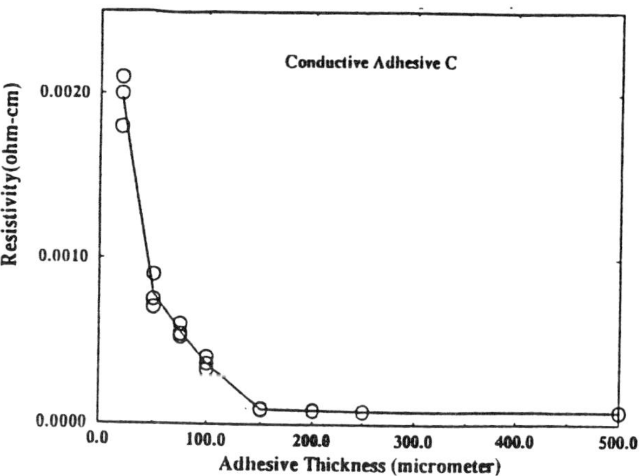

FIG. 14 RELATIONSHIP BETWEEN RESISTIVITY AND ADHESIVE FILM THICKNESS FOR CONDUCTIVE ADHESIVE C

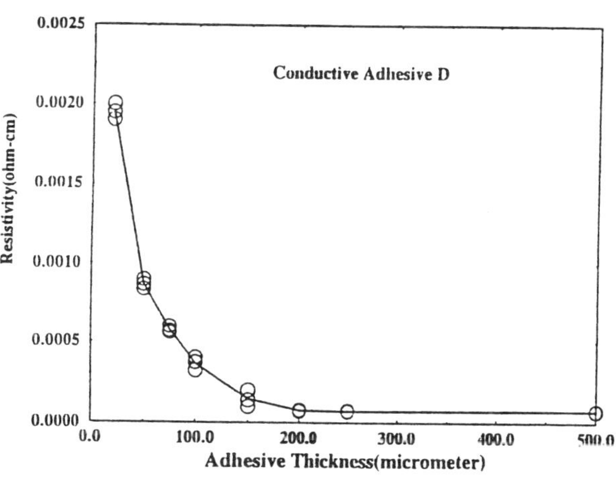

FIG. 15 RELATIONSHIP BETWEEN RESISTIVITY AND ADHESIVE FILM THICKNESS FOR CONDUCTIVE ADHESIVE D

**DE-Vol. 87, Reliability, Stress Analysis, and Failure Prevention
Issues in Emerging Technologies and Materials
ASME 1995**

ON THE ISSUE OF ELECTRIC BOUNDARY CONDITIONS
IN PIEZOELECTRIC DEFECT PROBLEMS

Horacio Sosa and Naum Khutoryansky
Department of Mechanical Engineering and Mechanics
Drexel University, Philadelphia PA 19104

ABSTRACT

The effects of the electric boundary conditions on the electro-elastic variables in the neighbourhood of cavities and cracks are analyzed within the realm of a two-dimensional, fully coupled theory of piezoelectricity. To this end, the case of an elliptic hole embedded in a transversely isotropic solid is considered. Comparisons between physically "correct" and so-called "impermeable" boundary conditions are provided for cavities of varying aspect ratios subjected to electrical and mechanical loads.

INTRODUCTION

The problem under consideration concerns a transversely isotropic piezoelectric solid containing an elliptic cavity with axes parallel to the axes of material symmetry. When the minor axis of the ellipse is made to vanish the cavity becomes a slit crack. The analysis of this configuration, or similar ones, has been addressed by many authors during the last few years, resorting not only to different methodologies but also to different sets of boundary conditions. Typical examples are furnished by the works of Deeg (1980), Dunn (1984), McMeeking (1989), Pak (1990, 1992a, 1992b), Shindo et al. (1990), Sosa and Pak (1990), Sosa (1991, 1992), Sosa and Khutoryansky (1995) and Suo et al. (1992).

Strictly speaking, the hole (which typically is filled with a gas like air) and the material around it are considered to be two dielectric mediums with different material constants. At the surface separating the two mediums (i.e., the rim of the cavity) there are no forces applied and no surface charges, thus in such a case the boundary conditions require the traction vector to vanish and the normal component of the induction vector and the electric potential to be continuous across the surface. Under these circumstances, the solution of a boundary-value problem requires analyzing both the interior and exterior of the hole, a process in most cases difficult to carry out in closed form. However, most man-made piezoelectric materials have dielectric constants 1000 to 4000 times larger than the dielectric constant of air, a condition that allows one to replace the aforementioned ("exact") electric boundary conditions by one stating that the normal component of the induction vector vanishes at the surface. In other words, the contour is assumed to be impermeable to the electric field (a condition similar to that of thermal insulation). Most mathematical models addressing fracture of piezoelectric materials are based on the approximate or impermeable condition because of the resulting mathematical simplifications: in such a case there is no need for modelling the interior of the hole.

In this article solutions for the aforementioned defect electro-elastic problem based on a model with exact boundary conditions (hereafter called EM) are provided in closed form. Moreover, results associated with the approximate or impermeable model (hereafter referred to as AM) are obtained from EM by simply setting to zero the dielectric constant of the gas enclosed by the hole.

An analysis of the behavior of the electric field within the hole, when a voltage or stress is applied to the solid at remote distances from the hole, is given for both models. It is shown that for relatively "open" holes, EM and AM yield virtually the same results. However, when the cavity tends to a slit crack, the prediction from these models diverge drastically: while EM yields large but bounded values for the field, AM predicts a singular electric field within the crack and, therefore, at its tip.

Finally, the influence of an applied electric field on the stresses induced in the material surrounding the hole is also considered. It is shown that, once again, while EM and AM yield indistinguishable results for wide open cavities, they provide completely different results when the ellipse becomes a mathematical crack.

MATHEMATICAL GENERALITIES

We consider a transversely isotropic solid, namely a crystal like quartz or ceramics that have been artificially polarized (Berlincourt et al., 1964 and van Randeraat and Setterington, 1974), described by the following set of constitutive equations:

$$S = s^D\, T + g^t\, D$$
$$E = -g\, T + \beta^T\, D \qquad (1)$$

where S, T, D and E are the strain, stress, induction (or electric displacement) and electric field, respectively. Moreover, s^D, g (with transpose g^t) and β^T are the compliance (measured at open circuit), the piezoelectric and impermeability (measured at constant stress) tensors.

Let the material described by (1) be polarized in the z-direction, that is (xy) is the corresponding basal isotropic plane. We are interested in a plane-strain problem such that full electromechanical effects can take place. Thus we consider the plane (xz) and assume that $S_{yy} = S_{yx} = S_{yz} = E_y = 0$. In such a case, after renaming the coordinates according to $x \to x_1$ and $z \to x_2$, Eq (1) becomes

$$S = a\, T + b^t\, D$$
$$E = -b\, T + c\, D \qquad (2)$$

where

$$S = \left\{ \begin{array}{c} S_{11} \\ S_{22} \\ 2S_{12} \end{array} \right\} ; \quad T = \left\{ \begin{array}{c} T_{11} \\ T_{22} \\ T_{12} \end{array} \right\}$$

$$D = \left\{ \begin{array}{c} D_1 \\ D_2 \end{array} \right\} ; \quad E = \left\{ \begin{array}{c} E_1 \\ E_2 \end{array} \right\}$$

and

$$a = \left\| \begin{array}{ccc} a_{11} & a_{12} & 0 \\ a_{12} & a_{22} & 0 \\ 0 & 0 & a_{33} \end{array} \right\|$$

$$b = \left\| \begin{array}{ccc} 0 & 0 & b_{13} \\ b_{21} & b_{22} & 0 \end{array} \right\| ; \quad c = \left\| \begin{array}{cc} c_{11} & 0 \\ 0 & c_{22} \end{array} \right\|$$

where

$$a_{11} = s_{11} - \frac{s_{12}^2}{s_{11}}, \quad a_{12} = s_{13} - \frac{s_{12}s_{13}}{s_{11}},$$

$$a_{22} = s_{33} - \frac{s_{13}^2}{s_{11}}, \quad a_{33} = s_{44},$$

$$b_{21} = (1 - \frac{s_{12}}{s_{11}})g_{31}, \quad b_{22} = g_{33} - \frac{s_{13}}{s_{11}}g_{31},$$

$$b_{13} = g_{15}, \quad c_{11} = \beta_{11}, \quad c_{22} = \beta_{33} + \frac{g_{31}^2}{s_{11}}$$

are the reduced or "effective" material constants.

The set of equations (2) is supplemented by the relations of elastic equilibrium and electrostatics, namely

$$\frac{\partial T_{11}}{\partial x_1} + \frac{\partial T_{12}}{\partial x_2} ; \quad \frac{\partial T_{12}}{\partial x_1} + \frac{\partial T_{22}}{\partial x_2} = 0$$

$$\frac{\partial D_1}{\partial x_1} + \frac{\partial D_2}{\partial x_2} = 0 \qquad (3)$$

and by the two-dimensinal relations of elastic compatibility and irrotationality of the electric field, that is

$$\frac{\partial^2 S_{11}}{\partial x_2^2} + \frac{\partial^2 S_{22}}{\partial x_1^2} - 2\frac{\partial^2 S_{12}}{\partial x_1 \partial x_2} = 0$$

$$\frac{\partial E_1}{\partial x_2} - \frac{\partial E_2}{\partial x_1} = 0 \qquad (4)$$

The solution to the plane problem can be found by introducing a stress function $U(x_1, x_2)$ and an induction function $\psi(x_1, x_2)$ such that (3) are satisfied if defined as

$$T_{11} = \frac{\partial^2 U}{\partial x_2^2} ; \quad T_{22} = \frac{\partial^2 U}{\partial x_1^2} ; \quad T_{12} = -\frac{\partial^2 U}{\partial x_1 \partial x_2} \qquad (5)$$

and

$$D_1 = \frac{\partial \psi}{\partial x_2} ; \quad D_2 = -\frac{\partial \psi}{\partial x_1} \qquad (6)$$

Moreover, combining (2) and (4)-(6) yields

$$L_4 U - L_3 \psi = 0$$
$$L_3 U + L_2 \psi = 0 \qquad (7)$$

where $L_i(i = 4, 3, 2)$ are elastic, piezoelectric, and dielectric differential operators of order four, three, and two, respectively, given by

$$L_4 = a_{22}\frac{\partial^4}{\partial x_1^4} + a_{11}\frac{\partial^4}{\partial x_2^4} + (2a_{12} + a_{33})\frac{\partial^4}{\partial x_1^2 \partial x_2^2}$$

$$L_3 = b_{22}\frac{\partial^3}{\partial x_1^3} + (b_{21} + b_{13})\frac{\partial^3}{\partial x_1 \partial x_2^2}$$

$$L_2 = c_{22}\frac{\partial^2}{\partial x_1^2} + c_{11}\frac{\partial^2}{\partial x_2^2}$$

The system given by (7) can be reduced to a single sixth order differential equation for U, which can be solved in terms of complex variables by letting $U(z) = U(x_1 + \mu x_2)$ where μ is a complex parameter. The introduction of this function in (7) yields an algebraic equation of order six with even powers in μ, which, consequently, can be solved in closed form yielding the roots μ_1, μ_2 and μ_3 and their corresponding complex conjugates. Hence, the functions U and ψ can be expressed as the sums of three functions U_k and ψ_k (one for each root) and their complex conjugates. However, rather than using these functions it is better to introduce the complex potentials $\varphi(z_k)$ defined as

$$\varphi_k(z_k) = U_k' = \frac{dU_k}{dz_k}; \quad z_k = x_1 + \mu_k x_2; \quad k = 1, 2, 3$$

by means of which the stress components are expressed in the form

$$\mathbf{T} = 2\Re \sum_{k=1}^{3} \begin{Bmatrix} \mu_k^2 \\ 1 \\ \mu_k \end{Bmatrix} \varphi_k' \tag{8}$$

where $\Re$ denotes the real part of the expression contained within the summation symbol, while the components of electric induction and field are given by

$$\mathbf{D} = 2\Re \sum_{k=1}^{3} \begin{Bmatrix} \mu_k \\ -1 \end{Bmatrix} \lambda_k \varphi_k'$$

$$\mathbf{E} = 2\Re \sum_{k=1}^{3} \begin{Bmatrix} 1 \\ \mu_k \end{Bmatrix} \kappa_k \varphi_k' \tag{9}$$

where

$$\lambda_k = -\frac{(b_{21} + b_{13})\mu_k^2 + b_{22}}{c_{11}\mu_k^2 + c_{22}} \tag{10}$$

and

$$\kappa_k = (b_{13} + c_{11}\lambda_k)\mu_k \tag{11}$$

The details involved in the steps leading to (8) and (9) are found in Sosa (1991).

The determination of the complex potentials is constrained to boundary conditions of mechanical and electrical type. Thus, if Γ_t, Γ_u, Γ_D and Γ_ϕ are the different parts of the body's boundary Γ where traction, displacement, induction and electric field, respectively, are prescribed we have in general (IEEE, 1988)

$$\mathbf{Tn} = \mathbf{t}^* \quad \text{on } \Gamma_t$$

$$\mathbf{u} = \mathbf{u}^* \quad \text{on } \Gamma_u$$

$$\mathbf{n} \cdot [\![\mathbf{D}]\!] = w_e \quad \text{on } \Gamma_D$$

$$[\![\phi]\!] = 0 \quad \text{on } \Gamma_\phi \tag{12}$$

with the conditions

$$\Gamma_t \cup \Gamma_u = \Gamma, \quad \Gamma_t \cap \Gamma_u = 0,$$

$$\Gamma_D \cup \Gamma_\phi = \Gamma, \quad \Gamma_D \cap \Gamma_\phi = 0$$

In Eq (12), $[\![f]\!] = f^+ - f^-$, where f^+ and f^- are the values of the function on opposite sides of the surface Γ. Moreover, w_e is a prescribed surface charge density and $\mathbf{n}$ is the outward unit normal to Γ. When dealing with a dielectric and its surroundings, usually vacuum or air, no surface charge will be present unless this one is placed there. Therefore, in the rest of the paper $w_e = 0$.

THE ELLIPTIC CAVITY PROBLEM

Let us now consider that the piezoelectric body contains an elliptic cavity (of contour denoted by Γ) with major and minnor axes $2a$ and $2b$ parallel to the material symmetry axes x_1 and x_2, respectively. The inside of the cavity is considered to be filled with a homogeneous gas[1] of dielectric constant (or permittivity) ϵ_0. The hole is assumed to be traction free, and as stated before no surface charge is considered or placed on the surface separating matter from vacuum. Mechanical and electrical loads in the form of forces, displacements, charge or voltages are applied to the body at remote distances from the cavity. The main purpose of the paper is to find expressions for the elastic and electric variables outside and inside the boundary Γ. In this last region, only electrical variables exist, to which we refer as $\mathbf{D}^c$ and $\mathbf{E}^c$. The latter is derived from the electric potential ϕ^c which satisfies Laplace's equation, that is

$$\nabla^2 \phi^c = 0 \tag{13}$$

---

[1] In most cases we can consider that the gas is simply air or perhaps vacuum in which case $\epsilon_0 = 8.85 \times 10^{-12}$ N/V^2.

In turn, the electric displacement is found via the constitutive relation

$$\mathbf{D}^c = \epsilon_0 \, \mathbf{E}^c$$

and its normal component is given by

$$\mathbf{D}^c \cdot \mathbf{n} = -\epsilon_0 \frac{\partial \phi^c}{\partial n}$$

Thus for the problem under consideration the boundary conditions on Γ become

$$\text{EM :} \quad \begin{cases} t_1 = 0 \\ t_2 = 0 \\ D_n = -\epsilon_0 \frac{\partial \phi^c}{\partial n} \\ \phi = \phi^c \end{cases} \qquad (14)$$

where the quantities to the left of the equal signs correspond to those evaluated within the piezoelectric body. The set (14) constitutes the basic ingredient of the EM model, from which the impermeable or AM model can be deduced by simply setting $\epsilon_0 = 0$. This last step is permissible in view of the fact that for most piezoelectric materials, their dielectric constants are much larger than those of air or vacuum. Therefore, an analysis based on AM requires the following set of boundary conditions:

$$\text{AM :} \quad \begin{cases} t_1 = 0 \\ t_2 = 0 \\ D_n = 0 \end{cases} \qquad (15)$$

The solution of the aforementioned problem is obtained by conformal mapping. Thus, interior and exterior regions to the hole are mapped into a circular ring, of external radius equal to one and inner radius of magnitude $[(a - b)/(a + b)]^{1/2}$. The mathematical details of obtaining the potential functions within the material are rather long and have been presented, very recently, elsewhere (Sosa and Khutoryansky, 1995).

All electro-mechanical variables can be determined within the material, which are found to be given by

$$\mathbf{T} = \mathbf{T}^\infty + 2\Re \sum_{k=1}^{3} \begin{Bmatrix} \mu_k^2 \\ 1 \\ \mu_k \end{Bmatrix} \mathcal{Z}_k$$

$$\mathbf{D} = \mathbf{D}^\infty + 2\Re \sum_{k=1}^{3} \begin{Bmatrix} \mu_k \\ -1 \end{Bmatrix} \lambda_k \mathcal{Z}_k \qquad (16)$$

$$\mathbf{E} = \mathbf{E}^\infty + 2\Re \sum_{k=1}^{3} \begin{Bmatrix} 1 \\ \mu_k \end{Bmatrix} \kappa_k \mathcal{Z}_k$$

where $\mathbf{T}^\infty$, $\mathbf{D}^\infty$ and $\mathbf{E}^\infty$ are the applied remote stress, induction and electric field, respectively,

$$\mathcal{Z}_k = \mathcal{Z}_k(\mu_k, z_k) = \frac{-a_{k1}}{\zeta_k^2 \omega_k'(\zeta_k)} \qquad (17)$$

and

$$z_k - \omega_k(\zeta_k) = \frac{a - i\mu_k b}{2}\zeta_k + \frac{a + i\mu_k b}{2}\frac{1}{\zeta_k} \qquad (18)$$

The parameters a_{k1} depend on the material properties and the applied loads, and their explicit expressions have been given also by Sosa and Khutoryansky (1995).

CHARACTERISTICS OF THE ELECTRIC FIELD INSIDE THE CAVITY

The expressions for the electric field inside the hole are found by solving (13) subject to (14). The most interesting result is that the field is uniform within the hole and its components are given by

$$E_1^c = E_1^\infty + \frac{2}{a}\sum_{k=1}^{3} \Re\{\kappa_k \, a_{k1}\}$$

$$E_2^c = E_2^\infty + \frac{2}{b}\sum_{k=1}^{3} \Im\{\kappa_k \, a_{k1}\} \qquad (19)$$

Once the applied load is known, the products $\kappa_k \, a_{k1}$ can be determined and their real and imaginary parts yield the final expression for $\mathbf{E}^c$. We consider now two problems of interest:

Problem 1: Let the remote applied load be given by $\mathbf{E}^\infty = E_1^\infty \, \mathbf{e}_1 + E_2^\infty \, \mathbf{e}_2$. In this case (19) becomes

$$E_1^c = \frac{(\gamma + 2\alpha) \, E_1^\infty}{\gamma + 2\alpha\epsilon_0 c_{11}}$$

$$E_2^c = \frac{(\alpha\gamma + 2c_{11}c_{22}^{-1}) \, E_2^\infty}{\alpha\gamma + 2\epsilon_0 c_{11}} \qquad (20)$$

where $\alpha = b/a$ and γ is a real number, typically in the range $1 < \gamma < 2$.

Notice that, if $E_1^\infty = 0$ and $E_2^\infty = E_0$, there is no field in the x_1-direction and

$$E_2^c = \frac{[\alpha\gamma + 2(c_{11}/c_{22})] \, E_0}{\alpha\gamma + 2\epsilon_0 c_{11}} \qquad (21)$$

Next, suppose that the material and aspect ratio of the hole are such that:

$$\epsilon_0 c_{11} << 1; \quad \alpha >> \epsilon_0 c_{11}; \quad \alpha < 1 \qquad (22)$$

Such conditions are easily fulfilled by any piezoelectric material containing holes with aspect ratios of say,

$1/100 < \alpha < 1$. As a result, Eq (20) becomes (to a very good approximation)

$$E_2^c \approx \left[1 + \frac{2(c_{11}/c_{22})}{\alpha\gamma}\right] E_0 \qquad (23)$$

which is the same result one would obtain from (20) by setting $\epsilon_0 = 0$. That is, for values of α small, but large compared with $\epsilon_0 c_{11}$, EM and AM yield the same result.

However, if $\alpha \sim \epsilon_0 c_{11}$ (which in practice implies a ratio of approximately $b/a \sim 10^{-4}$), Eq (20) produces

$$E_2^c \approx \frac{2(c_{11}/c_{22})E_0}{\alpha\gamma + \epsilon_0 c_{11}} \qquad (24)$$

since $c_{11}/c_{22} \sim 1$.

The result given by (24) indicates that EM and AM predict quite different qualitative and quantitative field characteristics when the ellipse tends to a slit crack. Indeed, from (24)

$$\lim_{\alpha \to 0} E_2^c \to \frac{2E_0}{\epsilon_0 c_{22}} \qquad (25)$$

that is, EM yields a large, but bounded value for the electric field everywhere in the crack, and therefore, at its tip. However, an analysis based on AM, which is obtained by first setting $\epsilon_0 = 0$ in (24), clearly produces a singular electric field as $\alpha \to 0$.

It is very important to note that since at, for example, the point (a,0) of Γ, there is continuity of the tangential component of the electric field, the value at that point for E_2^c is equal to the value at the same point of E_2 as calculated using $(16)_3$. That is, the electric field in the material at (a,0) is given by (24) and (25) whether the defect is a hole or a crack, respectively. Since for a ceramic, $\epsilon_0 c_{22} \sim 10^{-4}$, it is clear that in the presence of a crack, even small fields of 100 V/m can produce local depoling of the material, while larger values could produce dielectric breakdown .

Problem 2: Let the applied load be given by $\mathbf{T}^\infty = T_0\, \mathbf{e}_2 \otimes \mathbf{e}_2$, which induces within the cavity a uniform electric field with components in both directions given by

$$E_1^c = \frac{c_{11}\,[\gamma\,\Omega_R - 2\Lambda_I]\,T_0}{\gamma + 2\alpha\epsilon_0 c_{11}} \qquad (26)$$

and

$$E_2^c = \frac{c_{11}\,[2(b_{22}/c_{22}) + 2\Lambda_R + \gamma\,\Omega_I]\,T_0}{\alpha\gamma + 2\epsilon_0 c_{11}} \qquad (27)$$

where

$$\Lambda = \frac{\mu_2\lambda_1 - \mu_1\lambda_2}{\mu_2 - \mu_1}; \quad \Omega = \frac{\mu_1\mu_2(\lambda_2 - \lambda_1)}{\mu_2 - \mu_1}$$

and the subindices R and I stand for the "real" and "imaginary" parts of the above quantities.

In Eq (27), since $\alpha < 1$ and, in general, $\epsilon_0 c_{11} << 1$, we have

$$E_1^c \approx \left[\Omega_R - \frac{2\Lambda_I}{\gamma}\right] c_{11} T_0 \qquad (28)$$

That is, both models predict the same value for E_1^c, which in addition is independent of α.

The situation for Eq (27) is similar to that of (20). That is, under conditions (22), both models yield the same value of electric field, namely

$$E_2^c \approx \frac{c_{11}\,[2(b_{22}/c_{22}) + 2\Lambda_R + \gamma\Omega_I]\,T_0}{\alpha\gamma} \qquad (29)$$

However, when $\alpha \sim \epsilon_0 c_{11} << 1$, both terms in the denominator of (27) must remain and it is clear that EM and AM diverge in their predictions. Namely, when $\alpha \to 0$, EM gives large but bounded values of E_2^c. At this point we may inquire how large this value can be. It turns out that for ceramics, the terms in the numerator are of the same order of magnitude, approximately $\sim 10^{-10}$, which means that for the slit crack a tension T_0 induces a field $E_2^c \sim T_0 \times 10^{-2}$. Therefore, a stress of say 10^7 Pa, can induce a substantial electric field in the material surrounding the crack.

On the other hand, when in (27) $\epsilon_0 \to 0$ first, AM predicts a singular field as $\alpha \to 0$. Finally, the same conclusion is reached if the order of taking the limits is reversed. That is, first we can consider the impermeable model and then take the limit when the ellipse becomes a crack.

STRESSES INDUCED BY AN APPLIED ELECTRIC FIELD

One of the most interesting aspects concerning piezoelectric materials with cracks has to do with the effects the electric field has on the stress distribution around a crack tip. We note that (at least from a qualitative point of view), the fracture characteristics of cracked piezoelectric solids subjected to purely mechanical loads can be described by fracture mechanics concepts of anisotropic media (Sosa and Pak, 1990). That is, the asymptotic expressions for stresses have the classical singularity $1/\sqrt{r}$ at the tip of the crack and the functions reflecting the angular distributions are also functions of the material properties.

The case of the piezoelectric solid subject, in addition to mechanical forces, to a remote electric field has additional consequences that we would like to investigate briefly. It is interesting to note that some experimental observations and analytical models have speculated that an electric field may enhance or retard crack propagation initiated by the application of mechanical forces. In this section we study, therefore, the case of the piezoelectric solid subjected to a field $\mathbf{E}^\infty = E_0 \, \mathbf{e}_2$ and draw our attention to the behavior of the stress component T_{22} on the boundary of the hole, in particular, at the point $(a, 0)$. Eq $(16)_1$ gives

$$T_{22}(a, 0) = \frac{K \, (c_{22}^{-1} - \epsilon_0) \, E_0}{\alpha\gamma + 2\epsilon_0 c_{11}} \tag{30}$$

where K is a material parameter. If the piezoelectric solid is, in addition, subjected to remote forces, the corresponding stresses can be added to (30) according to the principle of superposition. On this point, it is useful to note that the sign of $T_{22}(a, 0)$ in (30) depends solely on the sign of E_0, since the rest of the expression is positive. Thus, according to (30), the electric field can increase or reduce the intensity of the normal stress generated by say tensile tractions applied in an independent manner.

Under (22) and since $c_{22}^{-1} >> \epsilon_0$ we have

$$T_{22}(a, 0) \approx \frac{K E_0}{c_{22}\alpha\gamma} \tag{31}$$

and, once again, the results deduced from EM and AM are virtually the same. To estimate the magnitude of the stress component we notice that for a ceramic, $K c_{22}^{-1} \sim 10^8 - 10^9$, $\gamma \sim 1$. Hence a field of 10^5 V/m applied alone could induce a stress of up to 100 MPa if the axes of the ellipse are in the ratio $1/100$.

There are, however, limitations regarding the approximate nature of the estimations made above. For very slender ellipses or blunt cracks $\alpha\gamma$ may become comparable to the product $\epsilon_0 c_{11}$. In such a case one needs to look at the two limiting conditions discussed before, namely:

$$\lim_{\alpha \to 0} T_{22}(a, 0) = \frac{K \, (c_{22}^{-1} - \epsilon_0) \, E_2^\infty}{2\epsilon_0 c_{11}} \tag{32}$$

and the stress is inversely proportional to the permittivity of the gas enclosed by the crack. It is clear that if the analysis were based on a model that neglects the interior of the hole ($\epsilon_0 \to 0$) the normal stress would be unbounded, a result predicted by the approximate model.

If we start first with the condition of electric impermeability, from (30) we obtain

$$\lim_{\epsilon_0 \to 0} T_{22}(0) = \frac{K \, c_{22}^{-1} \, E_2^\infty}{\alpha\gamma} \tag{33}$$

which again produces singular stresses when the hole becomes a crack.

CONCLUSIONS

The problem of an elliptic hole embedded in a transversely isotropic piezoelectric solid has been addressed within the framework of in-plane electro-elastic interactions and through the use of two different sets of electric boundary conditions.

It has been shown that while the approximate approach of neglecting the interior of the cavity yields results in agreement with a more physically realistic model for wide open holes, a drastic divergence in the prediction of the results is observed when the hole becomes a slit crack. Therefore, the nature of singular electric field and stress components at the tip of a crack is artificial and only a consequence of neglecting the interior of the hole.

ACKNOWLEDGMENT

This work was supported in part by the National Science Foundation under Grants Nos. MSS-9215296 and INT-9414656. Partial support was also provided by the Ministry of Education of Spain to H.S. during his visit to Instituto de Investigación Tecnológica, Universidad Pontificia Comillas, Madrid.

REFERENCES

Berlincourt, D. A., Curran, D. R. and Jaffe, H., 1964, "Piezoelectric and Piezoceramic Materials and their Function in Transducers." In: *Physical Acoustics*, Vol I-A, ed. by Mason, W. P., Academic Press.

Deeg, W. F., 1980, "The Analysis of Dislocation, Crack and Inclusion Problems in Piezoelectric Solids," Ph.D. Thesis, Stanford University, CA.

Dunn, M., 1994, "The Effects of Crack Face Boundary Conditions on the Fracture Mechanics of Piezoelectric Solids," *Eng. Fracture Mech*, Vol.48, pp. 25-39.

IEEE, 1988, *Standard on Piezoelectricity,* ANSI Std 176-1987, The Institute of Electrical and Electronics Engineers, Inc, New York.

Ikeda, T., 1990, *Fundamentals of Piezoelectricity*, Oxford University Press, Oxford.

Jaffe, B., Cook, W.R. and Jaffe, H., 1971, *Piezoelectric Ceramics*, Academic Press, New York.

McMeeking, R. M., 1989, "Electrostrictive Stresses Near Crack-Like Flaws," *ZAMP*, Vol. 40, pp. 615-627.

Pak, Y. E., 1990, "Crack Extension Force in a Piezoelectric Material," *ASME J. Appl. Mech.*, Vol. 112, pp. 647-653.

Pak, Y. E., 1992a, "Linear Electro-Elastic Fracture Mechanics of Piezoelectric Materials," *Int. J. Fracture*, Vol. 54, pp. 79-100.

Pak, Y. E., 1992b, "Circular Inclusion Problem in Antiplane Piezoelectricity," *Int. J. Solids Structures*, Vol. 29, pp. 2403-2419.

Shindo, Y., Ozawa, E. and Nowacki, J. P., 1990, "Singular Stress and Electric Fields of a Cracked Piezoelectric Strip," *Appl. Electromag. Mater.*, Vol. 1, pp. 77-87.

Sosa, H. A. and Pak, Y. E., 1990, "Three-Dimensional Eigenfunction Analysis of a Crack in a Piezoelectric Material," *Int. J. Solids Structures*, Vol. 26, pp. 1-15.

Sosa, H., 1991, "Plane Problems in Piezoelectric Media with Defects," *Int. J. Solids Structures*, Vol. 28, pp. 491-505.

Sosa, H., 1992, "On the Fracture Mechanics of Piezoelectric Solids," *Int. J. Solids Structures*, Vol. 29, pp. 2613-2622.

Sosa, H. and Khutoryansky, N., 1995, "New Developments Concerning Piezoelectric Materials with Defects. *Int. J. Solids Structures*, (to appear).

Suo, Z., Kuo, C.M., Barnett, D.M. and Willis, J.R., 1992, "Fracture Mechanics for Piezoelectric Ceramics," *J. Mech. Phys. Solids*, Vol. 40, pp. 739-765.

van Randeraat, J. and Setterington, R. E., 1974, *Piezoelectric Ceramics*, Second edition, Mullard House, New York.

**DE-Vol. 87, Reliability, Stress Analysis, and Failure Prevention
Issues in Emerging Technologies and Materials
ASME 1995**

Two-Dimensional Thermal Stress Analysis in Butt Adhesive Joints Containing Hole Defects and Rigid Fillers in Adhesive under Non-Uniform Temperature Field

Yuichi Nakano, Masahide Katsuo & Masataka Kawawaki
Department of Mechanical Engineering
Shonan Institute of Technology
Kanagawa, Japan

Toshiyuki Sawa
Department of Mechanical Engineering
Yamanashi University
Yamanashi, Japan

ABSTRACT

This study is concerned with the thermal stress analysis of a butt adhesive joint which contains circular holes and rigid fillers in the adhesive and is under non-uniform temperature field. In the analysis, the adherends are assumed to be rigid and the adhesive is replaced with a finite strip having holes and rigid fillers in it and the thermal stress distribution in the adhesive is analyzed using a two-dimensional theory of elasticity. The effects of size and location of rigid fillers and circular holes on the stress distributions at the interface and at the filler and hole peripheries are clarified by numerical calculations. For verification, photoelastic experiments are performed using an epoxide resin plate with small holes and fillers in it, modeled as an adhesive in the joint. The analytical results are fairly consistent with the experimental ones.

INTRODUCTION

Adhesive joints are now widely used in mechanical structures because of easy joining of different materials, lightening joined structure, uniform stress distribution and so on. However, a lot of difficulties still remain and should be overcome in order to apply adhesive joints at important parts of machine structures with sufficient reliability. Thermal strength of the adhesive joint is one of the important problems since mechanical and thermal properties of an adherend and an adhesive are generally quite different so that the thermal stress is generated easily in the joint. In practical joints, an adhesive often contains fillers of various kinds of materials or small air holes created in applying it on bonding surfaces. Recently, a few experimental researches (Sohn, 1985; Young, 1986, Jakusik et al.,1990) have been carried out on the improvement of static and dynamic strengths of adhesively bonded joints by adding fillers such as rubbery, alumina and metal particles into an adhesive. However, analytical studies concerning the thermal strength of adhesive joints which contain such kinds of fillers (Nakagawa et al.,1994) and hole defects (Nakano et al., 1992; Nakagawa et al.,1993) in an adhesive are few and adhesives are mostly assumed to contain no such hole defects or fillers.

The purpose of this study is to clarify the effects of both rigid fillers and circular holes contained in an adhesive on the thermal stress distribution in butt adhesive joints where the adherends are under different constant temperatures. In the analysis, the adherends are replaced with semi infinite strips and the adhesive is modeled as a finite strip with circular holes and rigid fillers in it. Firstly the steady state temperature distribution in an adhesive is calculated analytically and then the thermal stress distribution is analyzed using a two-dimensional theory of elasticity and an iteration method. The effects of size and location of the rigid fillers and circular holes on the thermal stress distributions at the interface between an adherend and an adhesive and at filler and hole peripheries are clarified by numerical calculations. In addition, the analytical results are compared with the photoelastic experimental results.

THEORETICAL ANALYSIS

Figure 1 shows an analytical model of a butt adhesive joint in which two similar adherends are joined by an adhesive having two circular holes of radius a at $\pm o_x$ on the x axis symmetrically with respect to the y axis and a filler of the same radius a at the center of an adhesive. The upper and lower adherends are kept at constant temperatures T_1 and T_2 respectively and heat transfers between the

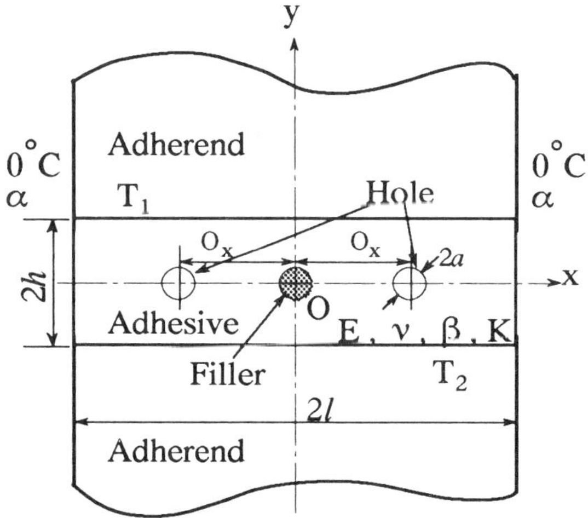

Fig.1 Model for analysis of butt adhesive joint having circular holes and a rigid filler in an adhesive

side surfaces of the joint and an ambient air of 0°C with a heat transfer coefficient α. In the analysis, the adhesive is replaced with a finite strip, the height and the width of which are denoted as $2h$ and $2l$, and Young's modulus, Poisson's ratio, coefficient of thermal expansion and thermal conductivity are E, ν, β and K, respectively. The following assumptions are made in this analysis: (1) The material properties are constant and independent of temperature change. (2) There is no temperature distribution in the z-direction of the joint, i.e., we have a two-dimensional heat transfer problem. (3) The adherends and fillers are rigid.

Temperature distribution

At first, the temperature distribution T(x,y) in an adhesive having no holes and fillers is analyzed. In the case of the steady state, temperature distribution is expressed as Eq.(1) and the thermal boundary conditions of the joint are given by Eq.(2):

$$\nabla^2 T(x,y) = 0 \tag{1}$$

$$T(x, +h) = T_1, \quad T(x, -h) = T_2, \quad -K\frac{\partial T(\pm l, y)}{\partial x} = \pm\alpha T(\pm l, y) \tag{2}$$

Eq.(1) can be solved and expressed as Eq.(3) using a variable separation method under the thermal conditions Eq.(2):

$$T(x,y) = \sum_{i=1}^{\infty} \{A_i \cosh(k_i y) + B_i \sinh(k_i y)\} \cos(k_i x) \tag{3}$$

$$A_i = \frac{2(T_1+T_2)\sin(k_i l)}{\cosh(k_i h)\{\sin(2k_i l)+2k_i l\}}, \quad B_i = \frac{2(T_1-T_2)\sin(k_i l)}{\sinh(k_i h)\{\sin(2k_i l)+2k_i l\}}$$

where, k_i is the i-th positive root satisfying $k_i \tan(k_i l) = $ h and h $= \alpha/K$.

Thermal stress

In this study, the thermoelastic potential approach is adopted for the thermal stress analysis. Generally, the thermoelastic potential $\Omega(x,y)$ is expressed by the temperature distribution T(x,y) as Eq.(4) and the thermal stresses σ_{x0}, σ_{y0}, τ_{xy0} and the displacements u_{x0} and v_{y0} in the x and y directions are obtained from $\Omega(x,y)$ as Eq.(5):

$$\nabla^2 \Omega(x,y) = T(x,y) \tag{4}$$

$$\sigma_{x0} = -E\beta\frac{\partial^2\Omega}{\partial y^2}, \quad \sigma_{y0} = -E\beta\frac{\partial^2\Omega}{\partial x^2}, \quad \tau_{xy0} = E\beta\frac{\partial^2\Omega}{\partial y^2}$$

$$2Gu_{x0} = E\beta\frac{\partial\Omega}{\partial x}, \quad 2Gv_{y0} = E\beta\frac{\partial\Omega}{\partial y} \tag{5}$$

Using Eq.(4), the thermoelastic potential $\Omega(x,y)$ for the adhesive is expressed as Eq.(6) and the thermal stresses and the displacements can be calculated using Eq.(5):

$$\Omega(x,y) = \sum_{i=1}^{\infty} \frac{1}{2k_i}\{A_i y\sinh(k_i y) + B_i y\cosh(k_i y)\} \cos(k_i x) \tag{6}$$

However, in practice, these solutions can not satisfy the boundary conditions of the joint concerning the stresses and the displacements which are expressed as Eq.(7), i.e., (a) at the interface between the adherend and the adhesive and at the side surfaces of the adhesive, (b) at a periphery of hole and (c) at a periphery of filler in the adhesive:

$$\sigma_x = \tau_{xy} = 0 \quad (x=\pm l), \qquad u_x = \frac{\partial v_y}{\partial x} = 0 \quad (y=\pm h) \cdots (a)$$

$$\sigma_r = \tau_{r\theta} = 0 \quad (r=a) \cdots (b), \quad u_r = v_\theta = 0 \quad (r=a) \cdots (c) \tag{7}$$

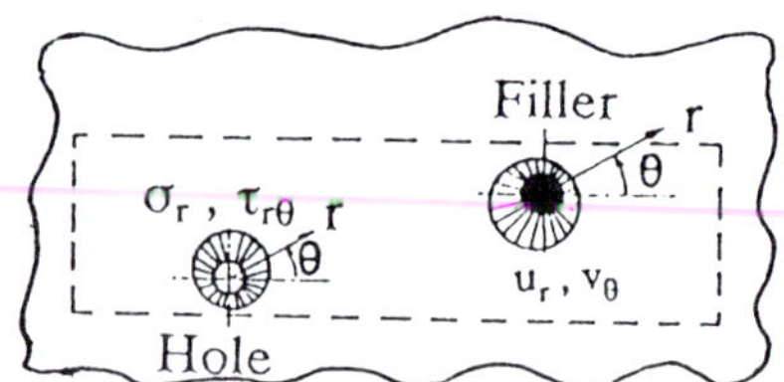

(a) Stresses at hole periphery and displacements at filler periphery

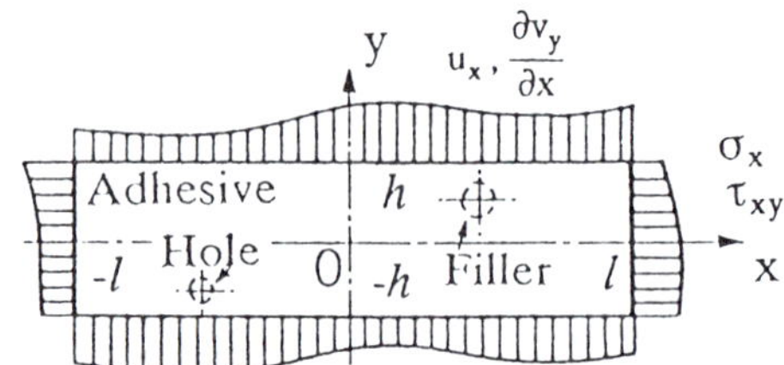

(b) Stresses and displacements at boundaries of adhesive

Fig.2 Illustration of analytical method

where, u_x and v_y are the displacements in the x and y directions, respectively, and u_r and v_θ are in the r and θ directions in polar coordinate, respectively. In order to satisfy the boundary conditions shown above, the following procedure is adopted in the present analysis.

Step 1: In order to satisfy the boundary conditions Eq.(7-a), the thermal stress distribution in the adhesive which has no holes and fillers, is solved using a two-dimensional theory of elasticity under the conditions of eliminating the stresses σ_{x0}, τ_{xy0} at the side surfaces and the displacements u_{x0}, $\partial v_{y0}/\partial x$ at the interface of the joint which are obtained from the thermoelastic potential $\Omega(x,y)$ (Nakano et.al.,1992). Then the stresses σ_r and $\tau_{r\theta}$ at each hole periphery (r = a) and the displacements u_r and v_θ at each filler periphery (r =a) are given analytically from this solution as shown in Fig.2(a).

Step 2: In order to satisfy the boundary condition Eq.(7-b), an infinite plate which has a circular hole is solved using Airy's stress function $\Psi(r,\theta)$ in polar coordinates expressed as Eq.(8) (Nakano et.al.,1988; Temma et al.,1991) under the condition of eliminating the stresses σ_r and $\tau_{r\theta}$ obtained in Step 1:

$$\Psi(r,\theta) = A_0\theta + B_0\log r$$
$$+ A_1 r \log r \cos\theta + B_1 r\theta \sin\theta + C_1 r\theta \cos\theta + D_1 r \log r \sin\theta$$
$$+ \sum_{n=2}^{\infty}(A_n r^{-n} + B_n r^{-n+2})\cos n\theta + \sum_{n=2}^{\infty}(C_n r^{-n} + D_n r^{-n+2})\sin n\theta \tag{8}$$

where, A_0, B_0, A_n, B_n, C_n and D_n (n=1,2,3,...) are unknown coefficients and can be determined comparing the coefficients in Fourier series of the negative stresses ($-\sigma_r$ and $-\tau_{r\theta}$) around each hole.

Similarly, an infinite plate in which a rigid filler is contained is analyzed using the similar stress function $\Psi(r,\theta)$, under the conditions of eliminating the displacements u_r and v_θ around each filler as shown in Fig.2(a). The displacements u_x and $\partial v_y/\partial x$ at the interfaces of the adhesive (y =$\pm h$) and the stresses σ_x and τ_{xy} at free side surfaces of the adhesive (x =$\pm l$) are obtained analytically from this solution.

Step 3: The adhesive is analyzed using Airy's stress function $\chi(x,y)$ which is expressed as Eq.(9) (Nakano et.al.,1988), under the conditions of eliminating the displacements and the stresses at the interfaces and the free side surfaces of the adhesive, respectively generated in Step 2 and shown in Fig.2(b):

$$\chi(x,y) = \chi_0 + \chi_1 + \chi_2 + \chi_3 + \chi_4 \qquad (9)$$

$$\chi_0 = \frac{A_0}{2}x^2 + \frac{B_0}{2}y^2 + \frac{C_0}{2}x^2 y$$

$$\chi_1 = \sum_{n=1}^{\infty} \frac{\overline{A}_n}{\overline{\Delta}_n \alpha_n^2}[\{\sinh(\alpha_n l)+\alpha_n l\cosh(\alpha_n l)\}\cosh(\alpha_n x)$$
$$-\sinh(\alpha_n l)\alpha_n x\sinh(\alpha_n x)]\cos(\alpha_n y)$$
$$+\sum_{s=1}^{\infty} \frac{\overline{B}_s}{\overline{\Omega}_s \lambda_s^2}[\{\sinh(\lambda_s h)+\lambda_s h\cosh(\lambda_s h)\}\cosh(\lambda_s y)$$
$$-\sinh(\lambda_s h)\lambda_s y\sinh(\lambda_s y)]\cos(\lambda_s x)$$

$$\chi_2 = \sum_{n=1}^{\infty} \frac{\overline{\overline{A}}_n}{\overline{\overline{\Delta}}_n \alpha_n'^2}[\{\sinh(\alpha_n' l)+\alpha_n' l\cosh(\alpha_n' l)\}\cosh(\alpha_n' x)$$
$$-\sinh(\alpha_n' l)\alpha_n' x\sinh(\alpha_n' x)]\sin(\alpha_n' y)$$
$$+\sum_{s=1}^{\infty} \frac{\overline{\overline{B}}_s}{\overline{\overline{\Omega}}_s \lambda_s^2}[\{\cosh(\lambda_s h)+\lambda_s h\sinh(\lambda_s h)\}\sinh(\lambda_s y)$$
$$-\cosh(\lambda_s h)\lambda_s y\cosh(\lambda_s y)]\cos(\lambda_s x)$$

$$\chi_3 = \sum_{n=1}^{\infty} \frac{\overline{A}_n'}{\overline{\Delta}_n' \alpha_n'^2}\{\cosh(\alpha_n' l)\alpha_n' x\sinh(\alpha_n' x)$$
$$-\alpha_n' l\sinh(\alpha_n' l)\cosh(\alpha_n' x)\}\cos(\alpha_n' y)$$
$$+\sum_{s=1}^{\infty} \frac{\overline{B}_s'}{\overline{\Omega}_s' \lambda_s'^2}\{\cosh(\lambda_s' h)\lambda_s' y\sinh(\lambda_s' y)$$
$$-\lambda_s' h\sinh(\lambda_s' h)\cosh(\lambda_s' y)\}\cos(\lambda_s' x)$$

$$\chi_4 = \sum_{n=1}^{\infty} \frac{\overline{\overline{A}}_n'}{\overline{\overline{\Delta}}_n' \alpha_n^2}\{\cosh(\alpha_n l)\alpha_n x\sinh(\alpha_n x)$$
$$-\alpha_n l\sinh(\alpha_n l)\cosh(\alpha_n x)\}\sin(\alpha_n y)$$
$$+\sum_{s=1}^{\infty} \frac{\overline{\overline{B}}_s'}{\overline{\overline{\Omega}}_s' \lambda_s'^2}\{\sinh(\lambda_s' h)\lambda_s' y\cosh(\lambda_s' y)$$
$$-\lambda_s' h\cosh(\lambda_s' h)\sinh(\lambda_s' y)\}\cos(\lambda_s' x)$$

where, $\alpha_n=\dfrac{n\pi}{h}$, $\alpha_n' = \dfrac{(2n-1)\pi}{2h}$, $\lambda_s=\dfrac{s\pi}{l}$, $\lambda_s' = \dfrac{(2s-1)\pi}{2l}$ (n, s =1,2,3...)

A_0, B_0, C_0, $\overline{A}_n$, $\overline{B}_s$,..., $\overline{\overline{B}}_s'$ (n, s =1,2,3,...) are unknown coefficients and can be determined from the boundary condition, Eq.(7-a). The displacements u_r and v_θ and the stresses σ_r and $\tau_{r\theta}$ are generated again, respectively, at each rigid filler and circular hole peripheries from this solution.

The Steps 2 and 3 are iterated until the boundary conditions Eq.(7) are satisfied sufficiently and finally the thermal stress and the displacement distributions in the adhesive can be obtained.

EXPERIMENTAL METHOD

Figure 3 shows the dimensions of a butt adhesive joint used in the photoelastic experiment. An epoxide resin plate which had three holes of 5 mm diameter at the center and at ±20mm from the center was used as an adhesive and structural steel (S45C, JIS) was used as an adherend and a filler. Young's modulus of the steel (206GPa) is about 60 times as large as that of the epoxide resin (3.43GPa) so that the adherends and fillers were considered to be relatively rigid.

Bonding surfaces of the steel and the epoxide plate were finished by grinding and bonded together at a constant temperature by an epoxide type adhesive (Sumitomo 3M, 1838B/A), material properties of which were similar to those of the epoxide plate. Simultaneously, two fillers were inserted into the holes on the

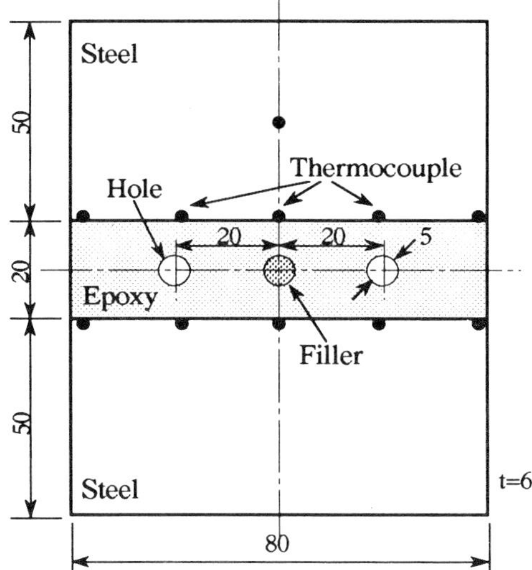

Fig.3 Dimensions of butt adhesive joint used in photoelastic experiment

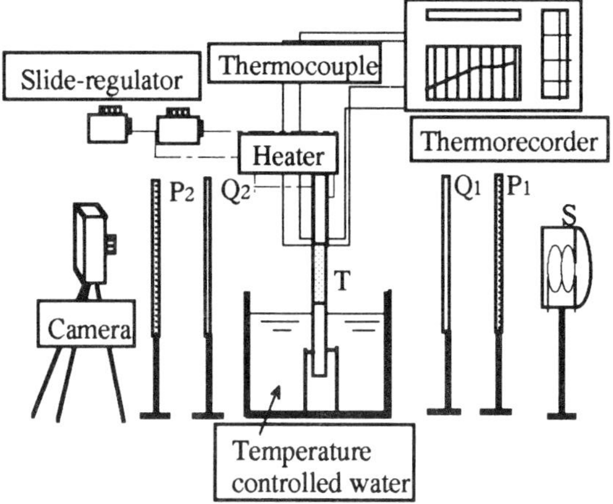

P$_1$,P$_2$: polarizers   Q$_1$,Q$_2$: quarter-wave plates
T : test specimen   S : light source
Fig.4 Experimental setup

epoxide plate and bonded by the same adhesive. Afterwards, the two adherends were kept at different temperatures, i.e., the upper one was heated by electric heaters attached on both front and back surfaces of it and the lower one was cooled in temperature controlled water as shown in Figure 4. Temperature distributions of each adherend were measured by thermocouples and when the joint was in a steady state temperature, isochromatic fringe pattern produced on the epoxide plate was observed by photoelasticity. It was confirmed in advance that an internal stress generated during joint manufacturing process was negligibly small over the entire epoxide plate from the observation of the isochromatic fringe pattern on the plate.

ANALYTICAL RESULTS

The effects of size and location of circular holes and rigid fillers in an adhesive on the thermal stress distributions at the interface

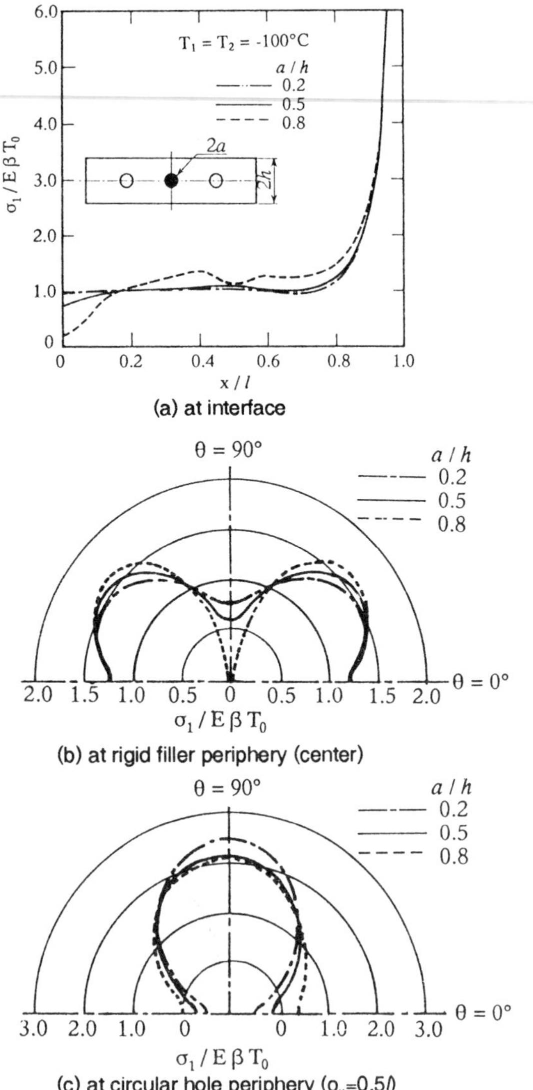

(a) at interface

(b) at rigid filler periphery (center)

(c) at circular hole periphery (o_x=0.5l)

Fig.5 Effect of ratio a/h on maximum principal thermal
stress distribution (joint (1))
(h/l=0.2, o_x/l=0.5, T_1=T_2=-100°C)

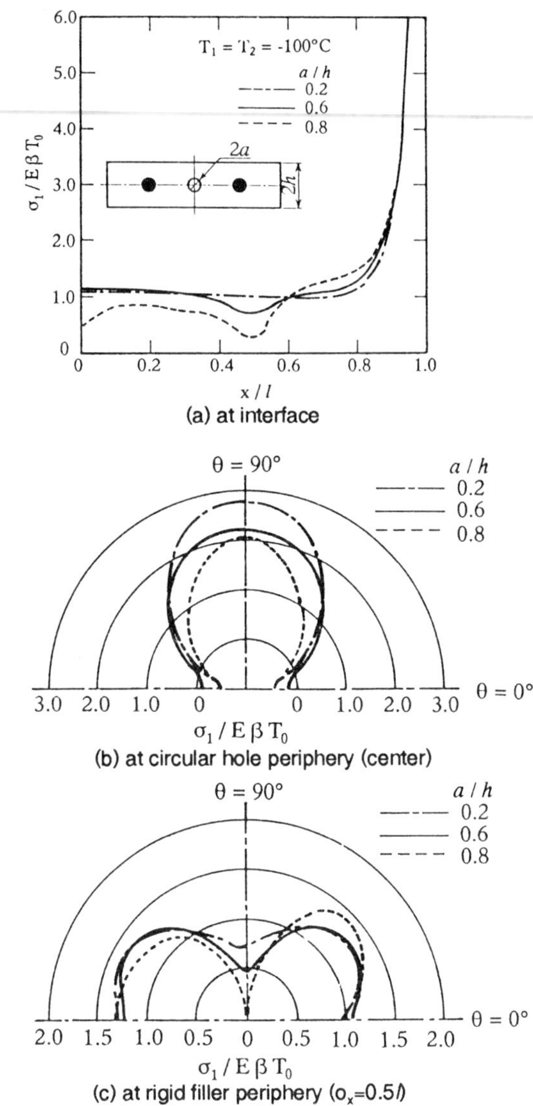

(a) at interface

(b) at circular hole periphery (center)

(c) at rigid filler periphery (o_x=0.5l)

Fig.6 Effect of ratio a/h on maximum principal thermal
stress distribution (joint (2))
(h/l=0.2, o_x/l=0.5, T_1=T_2=-100°C)

between an adherend and an adhesive and at hole and filler peripheries are examined numerically. In the numerical calculations, two kinds of butt adhesive joints are examined: the joint (1) where two circular holes are contained at $\pm o_x$ symmetrically with respect to the y axis and a rigid filler at the center of the adhesive, and the joint (2) where two rigid fillers and a circular hole are replaced vice versa. The number of terms N of the series in the analysis is taken as 80 and the number of iterations is 5. The difference in the thermal stresses of 5-th and 6-th iterations are within 1% in all calculations so that a satisfactory convergence is expected.

Figure 5 shows the effect of the ratio a/h on the maximum principal thermal stress distributions, (a) at the interface between the adherend and the adhesive, (b) at a rigid filler periphery located at the center and (c) at a circular hole periphery located at o_x=0.5l of the adhesive in the joint (1) when the upper and the lower adherends are kept at the same temperature (-100 °C). In the figures, the

maximum principal thermal stress is normalized as $\sigma_1/E\beta T_0$, where T_0 denotes the absolute temperature difference between the adherend and ambient air. The normalized maximum principal thermal stress distributions at a half interface of adhesion are shown in (a) and also those at half peripheries are shown in (b) and (c) based on the symmetry of the distribution. From (a), the normalized maximum principal thermal stress is tensile at the interface and is singular at the edge of it (x/l=1.0). The thermal stress around the center of the interface where a rigid filler is contained beneath decreases with an increase of radius a of a filler. The thermal stress around the interface where a circular hole is contained beneath also varies with hole size, however, on the contrary, it becomes slightly large with an increase of radius a. From (b), the normalized maximum principal stress at a filler periphery increases with an increase of filler size and the maximum value occurs at the positions in the directions θ of $\pm 45°$ and $\pm 135°$ of the center filler. On the

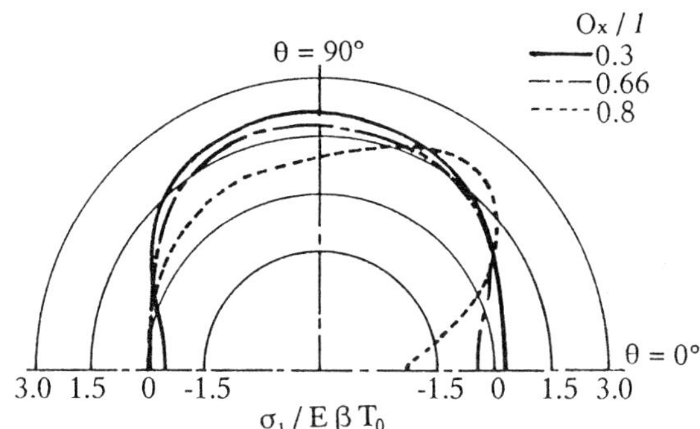

Fig.7 Effect of circular hole position on maximum principal stress distribution at hole periphery (joint (1)) (h/l=0.2, a/h=0.6, T_1=T_2=-100°C)

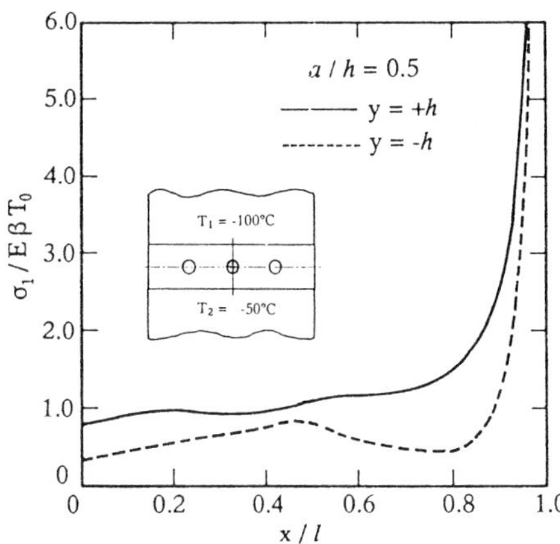

Fig.8 Effect of temperature of adherend on maximum principal stress distribution at interfaces (h/l=0.2, a/h=0.5, o_x/l=0.5, T_1=-100°C, T_2=-50°C)

other hand, from (c), the normalized maximum principal stress at a hole periphery increases with a decrease of hole size and the maximum value occurs around positions in the thickness direction (±90°) of the adhesive.

Similarly, Figure 6 shows the effect of the ratio a/h on the maximum principal thermal stress distributions in the joint (2), i.e., a hole is located at the center and two rigid fillers are at o_x=±0.5l. From (a), the normalized maximum principal thermal stress around the interface where a rigid filler is contained beneath decreases with an increase of filler size similarly to the joint (1) and that around the center of the interface seems to be less significant when hole radius a is smaller than 0.6h, but it decreases steeply when radius of a hole is larger as for a =0.8h. From (b) and (c), the effect of filler and hole size on the normalized maximum principal stresses at each periphery are similar to the joint (1), i.e., as radius a increases the maximum principal stress decreases at a hole periphery (b) and increases at a filler periphery (c) and from (c), the stress at the positions θ = ±45° is larger than that at θ = ±135° of a filler as its radius increases.

Figure 7 shows the effect of hole position on the normalized maximum principal stress distribution at a hole periphery in the joint (1) when radius a is 0.6h. The ratio o_x/l in the figure denotes the normalized distance between a filler located at the center and a hole in the adhesive. The maximum value occurs around the positions in the thickness direction (±90°) when a hole is located near the center of an adhesive (o_x/l =0.3) and it occurs in the direction θ of ±50° when located near the free side surface of an adhesive (o_x/l =0.8). The effect of hole position on the maximum principal stress distributions at the interface and at a filler periphery located at the center are, however, very small in this numerical condition. All the numerical results shown above are the case where two adherends undergo the same temperature change, i.e., T_1=T_2.

Figure 8 shows the case where two adherends are kept at different temperatures, i.e., an upper adherend is kept at T_1=-100°C and a lower one at T_2=-50°C in the joint (1). From the figure, the maximum principal thermal stress at the upper interface where the absolute value of the adherend temperature is higher becomes larger than that at the lower interface.

EXPERIMENTAL RESULTS

Figure 9 shows examples of isochromatic fringe patterns generated on the epoxide plate modeled as an adhesive of the joint (1). A rigid filler of 5 mm diameter is located at the center and two holes of the same size are at ±20mm from the center of the plate. The numerical results for the principal stress difference which coincides with the isochromatic lines are also shown for a half (right hand) side of the epoxide plate. Numbers shown in numerical results indicate orders of isochromatic line. (a) is the case when upper and lower adherends are kept at the same temperature (T_1=T_2=-20°C) and (b) is the case when upper and lower adherends are at T_1=-15°C and T_2=-20°C, respectively. Similarly, the experimental and numerical results for the joint (2) are shown in Figure 10. From both Figures 9 and 10, the isochromatic lines obtained by photoelastic experiments are symmetric with respect to the y axis of the joint and the thermal stress concentrates at filler and hole peripheries and near the edges of the interfaces. Moreover, the numerical results are fairly consistent with the experimental results in each case.

CONCLUSIONS

This study deals with the thermal stress analysis in a butt adhesive joint which has circular holes and rigid fillers in the adhesive. The joint has non-uniform temperature distribution, i.e., the upper and the lower adherends are kept at different temperatures. An epoxide plate is used to model the adhesive and the thermal stress distribution is measured by photoelastic experiment and compared with the analytical ones. The results obtained are as follows.

(1) A method to analyze the thermal stress distribution in a butt adhesive joint which contains circular holes and rigid fillers in the adhesive is demonstrated by using a two-dimensional theory of elasticity.

(2) The effects of size and location of a hole and a filler on the maximum principal thermal stress distributions at the interface and at hole and filler peripheries are clarified using numerical calculations.

(3) Photoelastic experiments were carried out in order to confirm the thermal stress analysis. The experimental results are fairly consistent with the numerical ones.

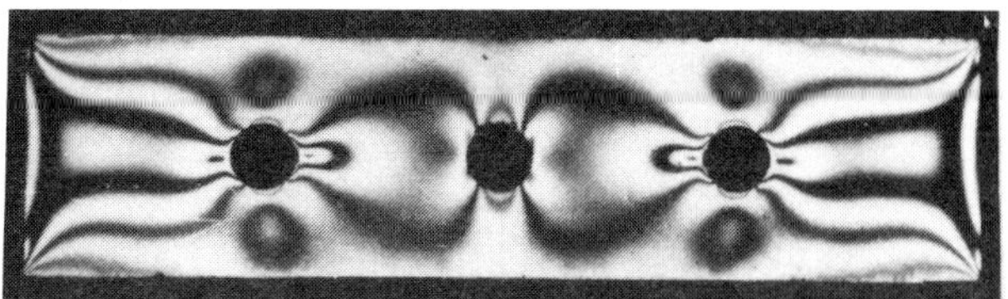

(a) The case when temperatures of upper and lower
adherends are -20°C

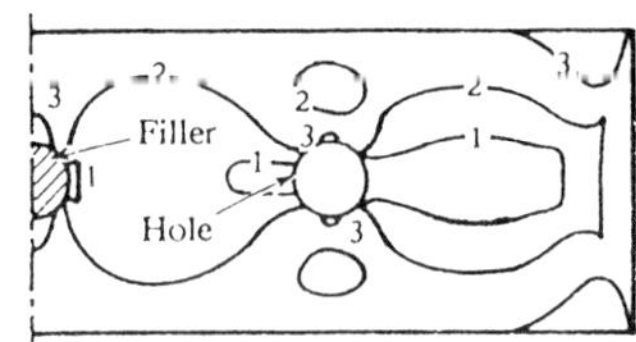

numerically obtained
isochromatic pattern

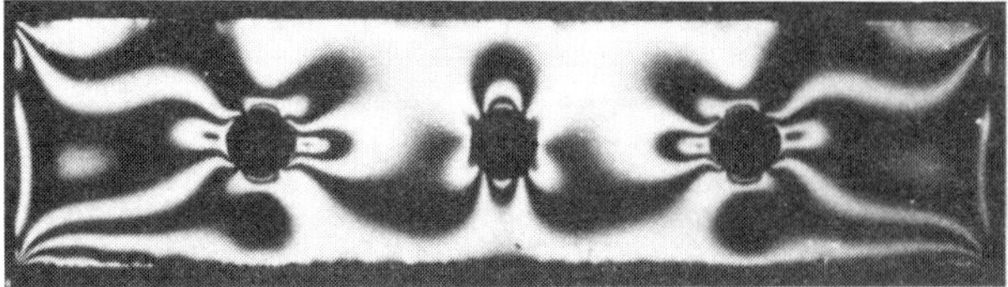

(b) The case when temperatures of upper and lower
adherends are -15°C and -20°C, respectively

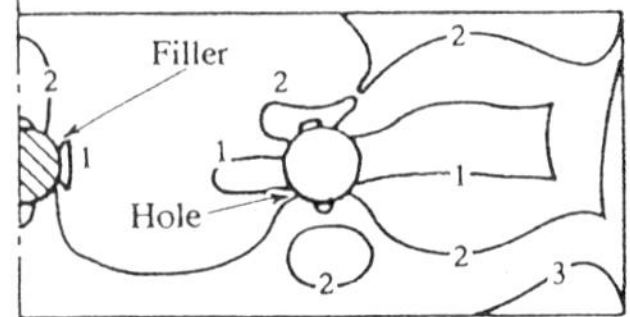

numerically obtained
isochromatic pattern

Fig.9 Photoelastic experimental results and comparison with numerical results　(joint (1))

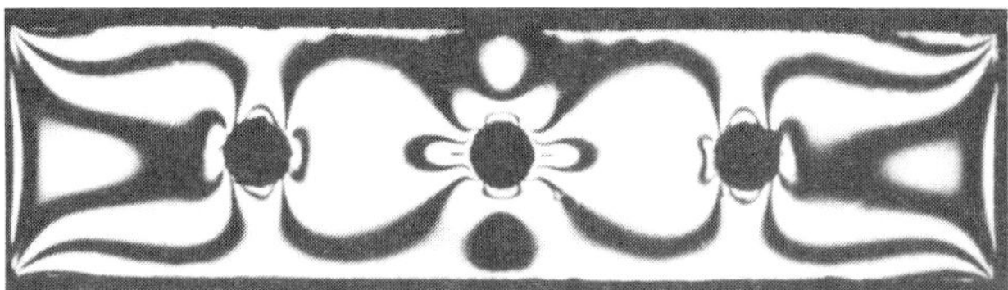

(a) The case when temperatures of upper and lower
adherends are -20°C

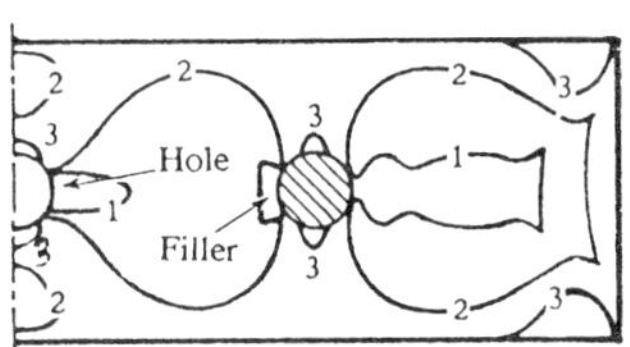

numerically obtained
isochromatic pattern

(b) The case when temperatures of upper and lower
adherends are -15°C and -20°C, respectively

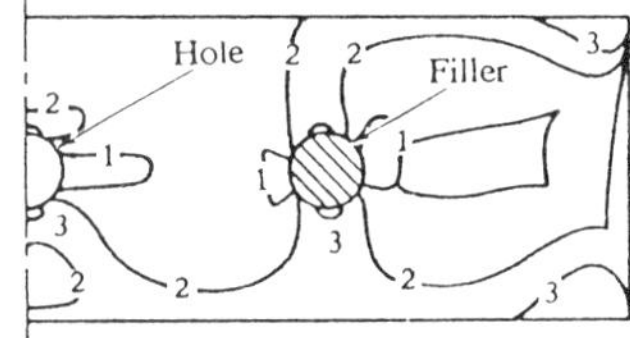

numerically obtained
isochromatic pattern

Fig.10 Photoelastic experimental results and comparison with numerical results　(joint (2))

REFERENCES

Jakusik,R., Jamarani, F, and Kinloch,A.J.,1990,"The Fracture Behavior of a Rubber-Modified Epoxy Under Impact Fatigue," *Journal of Adhesion*, Vol.32, pp.245-254.

Nakagawa,F., Sawa,T., Nakano,Y., and Hagiwara,S.,1993,"A Thermal Stress Analysis of Soldered/Bonded Joints with Defects," *Proceedings, ASME International Electronics Packaging Conference*, P.A.Engel and W.T.Chen ed., Vol.1, pp.49-54.

Nakagawa,F., Nakano,Y., and Sawa,T.,1994,"Two-Dimensional Thermal Stress Analysis of Butt Adhesive Joint Having Rigid Fillers in the Adhesive," *JSME International Journal*, Ser. A, Vol.37,No.3, pp.238-245.

Nakano,Y., Temma, K., and Sawa,T.,1988,"A Two-Dimensional Stress Analysis of Butt Adhesive Joints Having a Circular Fillers in the Adhesive," *JSME International Journal*, Ser. I, Vol.31, No.3, pp.507-513.

Nakano,Y., Nakagawa,F., and Sawa,T.,1992,"Photoelastic Measurement of Thermal Stress in Butt Adhesive Joints," *Proceedings, 7th International Congress on Experimental Mechanics*, The Society for Experimental Mechanics, Inc., Vol.1, pp.336-341.

Sohn,J.E.,1985,"Improved Matrix-Filler Adhesion," *Journal of Adhesion*, Vol.19, pp.15-27.

Temma,K., Sawa,T., Uchida, H., and Nakano,Y.,1991,"A Two-dimensional Stress Analysis of Butt Adhesive Joints Having a Circular Hole Defect in the Adhesive Subjected to External Bending Moments," *Journal of Adhesion*, Vol.33, pp.133-147.

Young,R.J.,1986, *Structural Adhesives*,Kinloch,A.J., ed., Elsevier Applied Science, pp.163-200.

**DE-Vol. 87, Reliability, Stress Analysis, and Failure Prevention
Issues in Emerging Technologies and Materials
ASME 1995**

TWO- DIMENSIONAL STRESS ANALYSIS AND STRENGTH EVALUATION
OF BAND ADHESIVE BUTT JOINTS SUBJECTED TO TENSILE LOADS

Toshiyuki SAWA
Department of Mechanical Engineering
Yamanashi University
Takeda, Kofu, Yamanashi
Japan

Yuichi NAKANO
Shonan Institute of Technology
Hujisawa, Kanagawa
Japan

Katsuhiro TEMMA
Kisarazu National College of Technology
Kisarazu, Chiba
Japan

Hiroaki UCHIDA
Kisarazu National College of Technology
Kisarazu, Chiba
Japan

ABSTRACT

The stress of band adhesive butt joints, in which two adherends were bonded partially at the interfaces, were analyzed, using a two−dimensional theory of elasticity , in order to demonstrate the usefulness of the joints. In the analysis, similar adherends and adhesive bonds, which were bonded at two or three regions, were respectively replaced by finite strips. In the numerical calculations, the effects of the ratio of Young's modulus for adherends to that for adhesives, the adhesive thickness, the bonding area and position, and the load distribution were shown on the stress distribution at the interface. It was seen that band adhesive joints were effective when the bonding area and positions were changed with external load distributions. Photoelastic experiments and experiments concerning the strains of adherends were conducted. The analytical results were in a fairly good agreement with the experimental results. In addition, a method for estimating joint strength is proposed by using the stress distribution at the interface obtained by the analysis. Experiments concerning the joint strength were performed. Fairly good agreement was found between the estimated values and the experimental results.

1. INTRODUCTION

Recently, adhesive joints have been used in mechanical structures, automobile industries and son as the performance of bonding materials has advanced. However, adhesive joints have not been used in the principal parts of mechanical structures, because the surface preparation is indeed, large deviations in the strength of the adhesive are caused and a non−defective method to examine the residual stress has not been established. Thus, it is expected that data will be made available in order to establish an optimal design method.

A lot of investigations have been carried out on butt, scarf and lap adhesive joints using the finite element method, (Baker,R.M. and Hatt,F., 1973, Sen,J.K. and Jones,R.M., 1980, Adams,R.D. and Harris,J.A., 1987, and Suzuki,Y, 1984) the photoelastic experiments (Chow,C.L. and Woo,C.W, 1984) and the theory of elasticity (Erdogan,F. and Ratwani,M., 1971, Wah,T., 1976, Renton,W.J., 1979, Sawa et al.,1987). However, the stress distribution of band adhesive joints, in which the adhesive efficiently resists external loads by bonding partially at the interface in comparison to the usual butt joints in which the adhesive are bonded completely at the interface, have not yet been fully elucidated, expect for some investigations (J.Pirvics, 1974). In addition, a band adhesive joint is expected to reduce the residual stress which will be caused in bonding process because the bonding area at the interface of the band adhesive joint is less than that of completely bonded adhesive joint. It is important to know the mechanical characteristics of band adhesive joints.

The objective of this study is to clarify the stress distribution at the interface and the effect of band bonding on the strength of butt adhesive joints when they are subjected to external tensile loads. By replacing similar adherends and adhesive by finite strips, the stress distribution of the joint is analyzed as a contact problem by using the two−dimensional theory of elasticity. The effect of the ratio of Young's modulus for adherends to that for adhesives, the adhesive thickness, the bonding area and position, and the external load distribution on the stress distribution at the interface between the adhesive and the adherends are clarified by the numerical calculations. In addition, a method for estimating the joint strength is proposed using the stress distribution. For verification, photoelastic experiments and experiments concerning the strains of adherends and the joint strength were conducted.   The numerical results were then compared with the experimental ones.

2. THEORETICAL ANALYSIS

Figure a illustrates a butt adhesive joint, in which two similar thin plates of finite length were bonded partially at the interface in the region of $|x_1| \leq l_2$ and $C-l_3 \leq |x_1| \leq C+l_3$, and subjected to an external tensile load.   In order to analyze the stress distributions at the interface between the adhesive and the adherends, two adherends were replaced by finite strip [I] and the adhesive with finite strips [II] and [III] as shown in Fig.2, taking into account the symmetry of the adhesive joint with respect to the x_2 and x_3 axes and the y_1 and y_2 axes of the adhesive.   The origins of each finite strip are designated as O_1, O_2, and O_3, respectively.   The length of the finite strip [I] is denoted by $2l_1$, the height by $2h_1$, Young's modulus by E_1, the shear modulus by G_1 and Poisson's ratio by ν_1.   Those of finite strips [II] and [III] are designated as $2l_2$, $2h_2$, E_2, G_2, ν_2, $2l_3$, $2h_2$, E_2, G_2 and ν_2, respectively. It is assumed that the tensile load acting on the upper and lower ends of the joint is replaced by a symmetrically distributed stress $F(x_1)$ with respect to the y_1 axis in the range of $e-d \leq |x_1| \leq e+d$ on the upper surface of finite strip [I].   Expanding the distribution $F(x_1)$ into the Fourier series, the boundary conditions shown in Fig.2 where the displacement in the x direction is denoted by u, the displacement in the y direction by v, and upper suffix I, II and III correspond to finite strips [I], [II] and [III], are expressed as follows;

(1) For finite strip [I] (adherends)

$$\left.\begin{aligned}
&x_1 = \pm l_1 : \sigma_x^I = \tau_{xy}^I = 0 \\
&y_1 = h_1 : \sigma_y^I = F(X) = a_0 + \sum_{s=1}^{\infty} a_s \cos\left(\frac{s\pi}{l}x\right) \\
&\tau_{xy}^I = 0
\end{aligned}\right\} \quad (1)$$

(2) For finite strip [I] (adhesive)

$$x_2 = \pm l_2 : \sigma_x^{III} = \tau_{xy}^{III} = 0 \quad (2)$$

(3) For finite strip [I] (adhesive)

$$x_3 = \pm l_3 : \sigma_x^{III} = \tau_{xy}^{III} = 0 \quad (3)$$

(4) At the boundary between finite strips [I] and [II]

$$\left.\begin{aligned}
&(\sigma_y^I)_{y_1=-h_1} = (\sigma_y^{II})_{y_2=h_2} \\
&(\tau_{xy}^I)_{y_1=-h_1} = (\tau_{xy}^{II})_{y_2=h_2} \\
&(u^I)_{y_1=-h_1} = (u^{II})_{y_2=h_2} \\
&\left(\frac{\partial v^I}{\partial x_1}\right)_{y_1=-h_1} = \left(\frac{\partial v^{II}}{\partial x_2}\right)_{y_2=h_2} \\
&(-l_2 \leq x_1 \leq l_2, -l_2 \leq x_2 \leq l_2)
\end{aligned}\right\} \quad (4)$$

(5) At the boundary between finite strips [I] and [III]

$$\left.\begin{aligned}
&(\sigma_y^I)_{y_1=-h_1} = (\sigma_y^{III})_{y_3=h_2} \\
&(\tau_{xy}^I)_{y_1=-h_1} = (\tau_{xy}^{III})_{y_3=h_2} \\
&(u^I)_{y_1=-h_1} = (u^{III})_{y_3=h_2} \\
&\left(\frac{\partial v^I}{\partial x_1}\right)_{y_1=-h_1} = \left(\frac{\partial v^{III}}{\partial x_3}\right)_{y_3=h_2} \\
&(-l_3 + C \leq |x_1| \leq l_3 + C, -l_3 \leq x_3 \leq l_3)
\end{aligned}\right\} \quad (5)$$

where

$$a_0 = \frac{1}{2l_1}\int_{-l_1}^{l_1} F(x)\,dx$$

$$a_s = \frac{1}{l_1}\int_{-l_1}^{l_1} F(x)\cos\left(\frac{s\pi}{l_1}x\right)dx \quad (s = 1,2,3,\ldots)$$

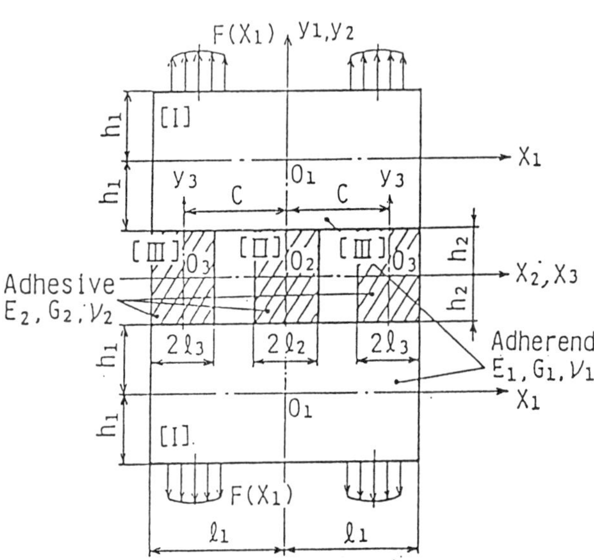

Fig.1 A band adhesive butt joint subjected to a tensile load

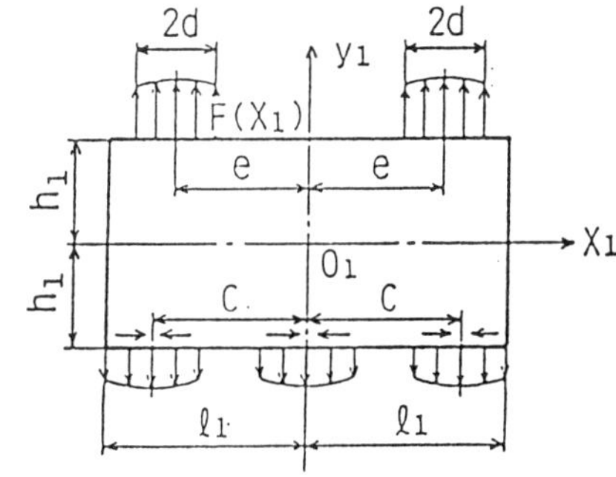

(a) finite strip [I] (adherend)

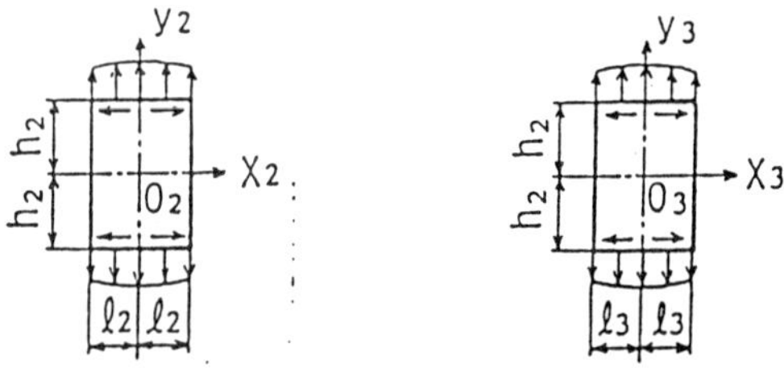

(b) finite strip [II] (adhesive) (c) finite strip [III] (adhesive)

Fig.2 A model for analysis and the dimensions

In order to analyze finite strips [I], [II], and [III] under the boundary conditions expressed by Eqs. (1)–(5), Airy's stress function χ is used.

Taking the boundary conditions into consideration, Aria's stress function χ^{I} which is used for the analysis of finite strip [I] is expressed as $\chi^{I} = \chi_1{}^{I} + \chi_2{}^{I} + \chi_3{}^{I} + \chi_4{}^{I}$ from solutions of variables separated of biharmonic functions (Sawa,T. et al, 1987, Sawa,T. et al., 1986), and the stress functions (Sawa,T. et al, 1987, Sawa,T. et al., 1986) χ^{II} and χ^{III} which are used for analysis of finite strips [II] and [III] are expressed as $\chi^{II} = \chi_1{}^{II} + \chi_3{}^{II}$ and $\chi^{III} = \chi_1{}^{III} + \chi_3{}^{III} + \chi_5{}^{III} + \chi_7{}^{III}$, respectively. The stress functions $\chi_1{}^{I}, \chi_2{}^{I}, \chi_3{}^{I}, \chi_4{}^{I}, \chi_1{}^{II}, \chi_3{}^{II}, \chi_1{}^{III}, \chi_3{}^{III}, \chi_5{}^{III}$ and $\chi_7{}^{III}$ are chosen as follows; where $A_0^{I}, \bar{A}_n^{I}, \bar{B}_s^{I}, \tilde{A}_n^{I}, \tilde{B}_s^{I}, \tilde{\bar{A}}_n^{I}, \tilde{\bar{B}}_s^{I}, A_0^{II}, \bar{A}_n^{II}, \bar{B}_s^{II}, \tilde{A}_n^{II}, \tilde{B}_s^{II}, A_0^{III}, A_n^{\bar{III}}, B_s^{\bar{III}}, A_n^{\bar{III}}, B_s^{\bar{III}}, A_n^{\bar{III}\prime}, B_s^{\bar{III}\prime}, A_n^{\bar{III}\prime}$ and $B_s^{\bar{III}\prime}$ in the equations are unknown coefficients which are determined from the boundary conditions.

$$\chi_1^{I} = \chi_1(A_0^{I}, \bar{A}_n^{I}, \bar{B}_s^{I}, \ell_1, h_1, \alpha_n^{I}, \lambda_s^{I}, \bar{\Delta}_n^{I}, \bar{\Omega}_s^{I}, x_1, y_1) \tag{6}$$

$$\chi_2^{I} = \chi_2(\bar{A}_n^{I}, \bar{B}_s^{I}, \ell_1, h_1, \alpha_n^{I\prime}, \lambda_s^{I}, \tilde{\bar{\Delta}}_n^{I}, \bar{\Omega}_s^{I}, x_1, y_1) \tag{7}$$

$$\chi_3^{I} = \chi_3(\tilde{A}_n^{I}, \tilde{B}_s^{I}, \ell_1, h_1, \alpha_n^{I\prime}, \lambda_s^{I\prime}, \tilde{\Delta}_n^{I}, \tilde{\Omega}_s^{I}, x_1, y_1) \tag{8}$$

$$\chi_4^{I} = \chi_4(\tilde{A}_n^{I}, \tilde{B}_s^{I}, \ell_1, h_1, \alpha_n^{I}, \lambda_s^{I}, \tilde{\bar{\Delta}}_n^{I}, \tilde{\bar{\Omega}}_s^{I}, x_1, y_1) \tag{9}$$

$$\chi_1^{II} = \chi_1(A_0^{II}, \bar{A}_n^{II}, \bar{B}_s^{II}, \ell_2, h_2, \alpha_n^{II}, \lambda_s^{II}, \bar{\Delta}_n^{II}, \bar{\Omega}_s^{II}, x_2, y_2) \tag{10}$$

$$\chi_3^{II} = \chi_3(\tilde{A}_n^{II}, \tilde{B}_s^{II}, \ell_2, h_2, \alpha_n^{II\prime}, \lambda_s^{II\prime}, \tilde{\Delta}_n^{II}, \tilde{\Omega}_s^{II}, x_2, y_2) \tag{11}$$

$$\chi_1^{III} = \chi_1(A_0^{III}, A_n^{\bar{III}}, B_s^{\bar{III}}, \ell_3, h_3, \alpha_n^{III}, \lambda_s^{III}, \Delta_n^{\bar{III}}, \Omega_s^{\bar{III}}, x_3, y_3) \tag{12}$$

$$\chi_3^{III} = \chi_3(A_n^{\tilde{III}}, B_s^{\tilde{III}}, \ell_3, h_3, \alpha_n^{III\prime}, \lambda_s^{III\prime}, \Delta_n^{\tilde{III}}, \Omega_s^{\tilde{III}}, x_3, y_3) \tag{13}$$

$$
\begin{aligned}
\chi_5^{I} &= \chi_5(A_n^{\bar{III}\prime}, B_s^{\bar{III}\prime}, \ell_3, h_2, \alpha_n^{III}, \lambda_s^{III\prime}, \Delta_n^{\bar{III}\prime}, \Omega_s^{\bar{III}\prime}, x_3, y_3) \\
&= \sum_{n=1}^{\infty} \frac{A_n^{III\,2}}{\Delta_n^{\bar{III}\prime} \alpha_n^{III\,2}} \big[\{\alpha_n^{III}\ell_3 \sinh(\alpha_n^{III}\ell_3) + \cosh(\alpha_n^{III}\ell_3)\} \sinh(\alpha_n^{III} x_3) \\
&\qquad - \alpha_n^{III} x_3 \cosh(\alpha_n^{III}\ell_3) \cosh(\alpha_n^{III} x_3)\big] \cos(\alpha_n^{III} y_3) \\
&\quad + \sum_{s=1}^{\infty} \frac{B_s^{\bar{III}\prime}}{\Omega_s^{\bar{III}\prime} \lambda_s^{III\prime\,2}} \big[\{\lambda_s^{III\prime} h_2 \cosh(\lambda_s^{III\prime} h_2) + \sinh(\lambda_s^{III\prime} h_2)\} \cosh(\lambda_s^{III\prime} y_3) \\
&\qquad - \lambda_s^{III\prime} y_3 \sinh(\lambda_s^{III\prime} h_2) \sinh(\lambda_s^{III\prime} y_3)\big] \sin(\lambda_s^{III\prime} x_3)
\end{aligned} \tag{14}
$$

$$
\begin{aligned}
\chi_7^{III} &= \chi_7(A_n^{\tilde{III}\prime}, B_s^{\tilde{III}\prime}, \ell_3, h_3, \alpha_n^{III\prime}, \lambda_s^{III}, \Delta_n^{\tilde{III}\prime}, \Omega_s^{\tilde{III}\prime}, x_3, y_3) \\
&= -\sum_{n=1}^{\infty} \frac{A_n^{\tilde{III}\prime}}{\Delta_n^{\tilde{III}\prime} \alpha_n^{III\prime\,2}} \big[\alpha_n^{III\prime}\ell_3 \cosh(\alpha_n^{III\prime}\ell_3) \sinh(\alpha_n^{III\prime} x_3) \\
&\qquad - \alpha_n^{III\prime} x_3 \sinh(\alpha_n^{III\prime}\ell_3) \cosh(\alpha_n^{III\prime} x_3)\big] \cos(\alpha_n^{III\prime} y_3) \\
&\quad - \sum_{s=1}^{\infty} \frac{B_s^{\tilde{III}\prime}}{\Omega_s^{\tilde{III}\prime} \lambda_s^{III\,2}} \big[\lambda_s^{III} h_2 \sinh(\lambda_s^{III} h_2) \cosh(\lambda_s^{III} y_3) \\
&\qquad - \lambda_s^{III} y_3 \cosh(\lambda_s^{III} h_2) \sinh(\lambda_s^{III} y_3)\big] \sin(\lambda_s^{III} x3)
\end{aligned} \tag{15}
$$

where

$$\alpha_n^{I} = \alpha_n(h_1) = \frac{n\pi}{h_1}, \quad \alpha_n^{I\prime} = \alpha_n{}'(h_1) = \frac{(2n-1)\pi}{2h_1}$$

$$\lambda_s^{I} = \lambda_s(\ell_1) = \frac{s\pi}{\ell_1}, \quad \lambda_s^{I\prime} = \lambda_s{}'(\ell_1) = \frac{(2s-1)\pi}{2\ell_1}$$

$$\alpha_n^{II} = \alpha_n(h_2) = \alpha_n^{III}, \quad \alpha_n^{II\prime} = \alpha_n{}'(h_2) = \alpha_n^{III\prime}, \quad \lambda_s^{II} = \lambda_s(\ell_2)$$

$$\lambda_s^{II\prime} = \lambda_s{}'(\ell_2), \quad \lambda_s^{III\prime} = \lambda_s{}'(\ell_3)$$

$$\bar{\Delta}_n^{I} = \Delta_n(\alpha_n^{I}\ell_1) = \sinh(\alpha_n^{I}\ell_1) \cosh(\alpha_n^{I}\ell_1) + \alpha_n^{I}\ell_1$$

$$\tilde{\bar{\Delta}}_n^{I} = \tilde{\Delta}_n^{I} = \Delta_n(\alpha_n^{I\prime}\ell_1), \quad \tilde{\Delta}_n^{I} = \bar{\Delta}_n^{I}, \quad \bar{\Delta}_n^{II} = \Delta_n(\alpha_n^{II}\ell_2)$$

$$\tilde{\Delta}_n^{II} = \Delta_n(\alpha_n^{II\prime}\ell_2), \quad \Delta_n^{\bar{III}} = \Delta_n(\alpha_n^{III}\ell_3), \quad \Delta_n^{\tilde{III}} = \Delta_n(\alpha_n^{III\prime}\ell_3)$$

$$\Delta_n^{\bar{III}\prime} = \Delta_n{}'(\alpha_n^{III}\ell_3) = \sinh(\alpha_n^{III}\ell_3) \cosh(\alpha_n^{III}\ell_3) - \alpha_n^{III}\ell_3$$

$$\Delta_n^{\tilde{III}\prime} = \Delta_n{}'(\alpha_n^{III\prime}\ell_3)$$

$$\bar{\Omega}_s^{I} = \Omega_s(\lambda_s^{I} h_1) = \sinh(\lambda_s^{I} h_1) \cosh(\lambda_s^{I} h_1) + \lambda_s^{I} h_1$$

$$\tilde{\bar{\Omega}}_s^{I} = \Omega_s(\lambda_s^{I} h_1) = \sinh(\lambda_s^{I} h_1) \cosh(\lambda_s^{I} h_1) - \lambda_s^{I} h_1$$

$$\tilde{\Omega}_s^{I} = \Omega_s(\lambda_s^{I\prime} h_1), \quad \tilde{\bar{\Omega}}_s^{I} = \Omega_s(\lambda_s^{I\prime} h_1), \quad \bar{\Omega}_s^{II} = \Omega_s(\lambda_s^{II} h_2)$$

$$\tilde{\Omega}_s^{II} = \Omega_s(\lambda_s^{II\prime} h_2), \quad \Omega_s^{\bar{III}} = \Omega_s(\lambda_s^{III} h_2) = \Omega_s^{\bar{III}\prime}, \quad \Omega_s^{\tilde{III}} = \Omega_s(\lambda_s^{III\prime} h_2) = \Omega_s^{\bar{III}\prime}$$

Using Airy's stress function χ^{I}, χ^{II} and χ^{III}, the stresses and the displacements of finite strips [I], [II] and [III] are obtained. By applying them to the boundary conditions, the relationships among the unknown coefficients are obtained. By solving an infinite set of simultaneous equations, the unknown coefficients $\bar{A}_n^{I}, \bar{B}_s^{I}, \ldots, A_n^{\bar{III}\prime}$ and $B_s^{\bar{III}\prime}$ are determined. However, the unknown coefficients A_0^{I}, A_0^{II} and A_0^{III} in Eq.(18) (see Appendix) cannot be determined immediately because this problem is statically intermediate. Thus, taking the equilibrium condition of the displacement in they direction into consideration, these coefficients, the stresses and the displacements are obtained.

3. EXPERIMENTAL METHOD

Figure 3 shows the dimensions of the specimen used in the experiments. Figure.3(a) shows the specimen used in the experiments concerning the strains of adherends. In the

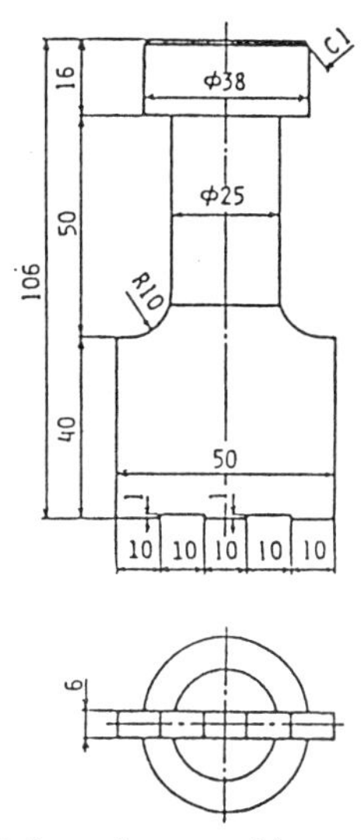

(a) A specimen used in experiments concerning the strains

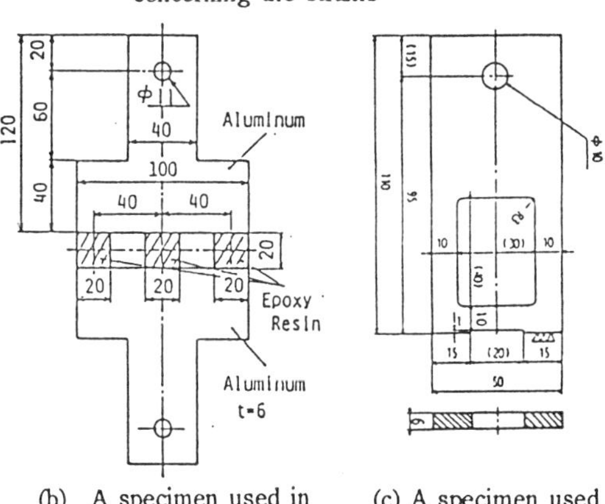

(b) A specimen used in photoelastic experiments

(c) A specimen used in joint strength tests

Fig.3 Dimensions of specimen used in experiments

interface of the specimen steps were manufactured to the depth of 1mm in order to make a disbonded area, and three bonded areas were made. The specimen is made of steel (S25C(JIS)), and the strain gauges were attached to the position of 5mm away from the interface. After bonding and solidifying a pair of specimens by using an epoxy bond (SUMITOMO 3M Co., Ltd., Scotch−Weld 1838) at room temperature, a tensile load was applied. The trains of the adherends were then measured. Figure (3) shows the specimen used in the photoelastic experiments. The adhesive was made of a type of epoxy resin and the adherend was made of aluminum. The thickness of the specimen is chosen as 6mm. The bonded surfaces were finished by lapping. After bonding and curing, photoelastic experiments were performed by applying a tensile load of 588N to the joint, and the stresses in the adhesive were measured. Figure 3(c) is dimensions of the specimens used in the experiments concerning the joint strength. The geometry of the specimens is designed in order to demonstrate the effectiveness of band adhesive joints. The specimens were made of steel (St) and aluminum (Al). The specimens of which the interface was completely contact were manufactured. A tensile load was applied to joint and the rupture load was measured.

4. NUMERICAL RESULTS AND COMPARISON BETWEEN THE NUMERICAL AND THE EXPERIMENTAL RESULTS

4.1 Numerical results

In the numerical calculations, stress singularity caused at the edge of the interface (y_2, $y_3 = \pm h_2$, $x_2 = \pm l_2$, $x_3 = \pm l_3$), so 50 terms of the series were chosen taking into account the convergence of the stresses because the difference of the calculations in each stress at the position of 0.004l_1 inside the disbonded part was smaller than 5% when we calculated by 50 and 55 terms of the series. Figure 4 shows the effect of the ratio E1/E2 of Young's modulus of the adherend to that of the adhesive on the each stress distribution at the interface ($y_1 = -h_1$, $y_2 = y_3 = \pm h_2$). Figure 4(a) shows the stress σ_y, σ_x and τ_{xy} distributions, and Fig. 4(b) shows the maximum principal stress distribution. The value of E_1/E_2 was chosen as 3, 10 and 100. It was assumed that the tensile load acts uniformly on the upper surface ($y_1 = h_1$) of the adherend (in Fig. 2, the distribution $F(x_1)$ is constant on $|x_1| \leq l_1$), and the bonded regions are put as $l_2 = l_3 = 0.2l_1$ and $C = 0.8l_1$. The ordinate is normalized each stress, where σ_{ym} is the mean normal stress of the adhesives [II] and [III] in the y_2 and y_3 directions. The abscissa is the ratio x_1/l_2 of the distance x_1 from the center of the adherend to the half length l_1 of the adherend. The closer the ratio E_1/E_2 approaches 1, the larger the stress singularity becomes at the edge of the interface, and the distribution of the maximum

principal stress σ_1 becomes to be more uniform with an increase of the ratio E_1/E_2.

Figure 5 shows the effect of the adhesive thickness on the maximum principal stress distribution at the interface. The ratio of h_1/h_2 of the height 2h_1 of the adherend to the thickness 2h_2 of the adhesive is chosen as 10, 25 and 100. The stress singularity at the edge of the interface decreases with an increase of the value of h_1/h_2.

Figure 6 shows the effect of bonding regions and their position on the maximum principal stress distribution at the interface, where the bonding area is held constant as 0.6l_1.

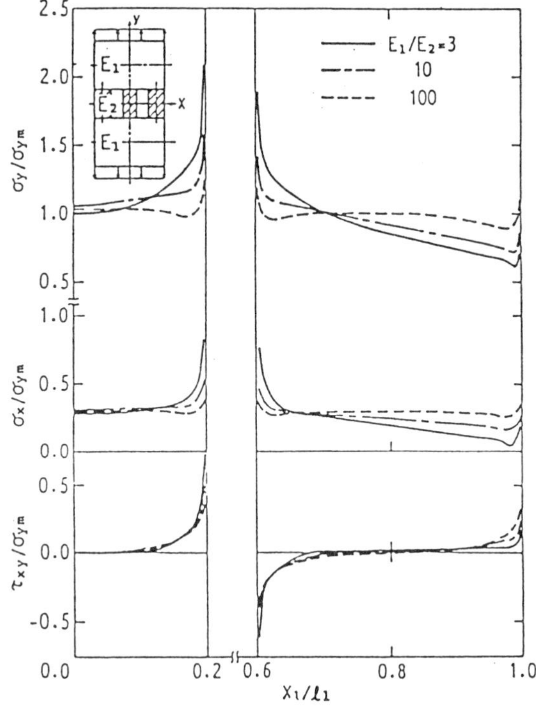

(a) each stress

(b) maximum principal stress

$$\left(\begin{array}{l} y_2=y_3=h_2,\ h_1/h_2=25,\ h_1/l_1=1,\ \nu_1/\nu_2=1, \\ l_2=l_3=0.2l_1,\ C=0.8l_1,\ \text{On } |x_1| \leq l_1,\ F(x_1) \\ \text{is constant} \end{array}\right)$$

Fig.4 An effect of ratio of Young's modulus on stress distribution at the interface

The bonding regions are chosen as the following three types, (i) one region at the center ($|x_1| \leq 0.6l_1$) of the joint, (ii) two regions at both ends ($0.4l_1 \leq |x_1| \leq l_1$) and (iii) three regions at the center and both ends ($|x_1| \leq 0.2l_1$, $0.6l_1 \leq |x_1| \leq l_1$). Each result is denoted in the figure by the symbols 1, 2 and 3, respectively. It is shown that the stress singularity near the edge of the interface decreases with an increase of the bonding region.

Figure 7 shows the maximum principal stresses at the points of $x_2 = 0.98l_2$, $x_3 = -0.98l_3$ and $x_3 = 0.98l_3$, when the adhesive [II] is fixed and the position of the adhesive [III] at both ends (the value C in Fig. 1) is changed. The values are denoted by A, B and C, respectively. When the value C (the distance to the origin O_3 shown in Fig. 1) becomes $0.68l_1$, the value of the maximum principal stresses A, B and C become approximately the same.

Figure 8 shows the effect of the distribution $F(x_1)$ of a tensile load on the maximum principal stress distribution at the interface of the band adhesive joint in the case of $h_1/l_2 = 0.2$. In the figure, the ordinate is the normalized maximum principal stress σ_1 / σ_{ym1}, where σ_{ym1} is the mean normal stress of the adherend [I] in the y_1 direction. Figure 8(a) is the case where the load distribution is uniform in the range $0.8l_1 \leq |x_1| \leq l_1$, the bonding region is two at both ends, the value of C is chosen ad $c = l_1 = l_3$ and the ratio of $2l_3/l_1$ is chosen 1.0 (bonding completely at the interface), 0.4 and 0.2. The values of σ_1 / σ_{ym1} is 6.69 at the point of $x_3 = -0.96l_3$. Thus, it is seen that the stress distribution in the case where the value of $2l_3/l_1$ is 0.4 is similar to the case where the value of $2l_3/l_1$ is 1.0 (bonded completely at the interface).

Figure 8(b) is the case where the distribution is uniform in the ranges $|x_1| \leq 0.05l_1$ and $0.9l_1 \leq |x_1| \leq l_1$, the ratio of $(l_2 + 2l_3)/l_1$ of bonding area ($l_2 + 2l_3$) to the length of adherend l_1 is chosen as 1.0, 0.6

and 0.3. In the case where the value of $(l_2 + 2l_3)/l_1$ is 0.3, the value of σ_1 / σ_{ym1} is 4.38 at the point of $x_3 = -0.96l_3$. However, in the case where the values of $(l_2 + 2l_3)/l_1$ are 0.6 and 1.0 (bonded completely at the interface), they have a similar stress distribution, the value of σ_1 / σ_{ym1} at the point of $0.99l_1$ are 3.47 and 3.55, respectively. Thus, in the case where the effect of load distribution occurs at the interface, it is expected that a band adhesive joint in which the interface is bonded partially has the same stress distribution as the case where the interface is bonded completely.

Figure 9 shows the relationship between the normalized maximum principal stress σ_1 / σ_{ym} and the distance r from the each edge, where x_1 / l_1 are chosen as

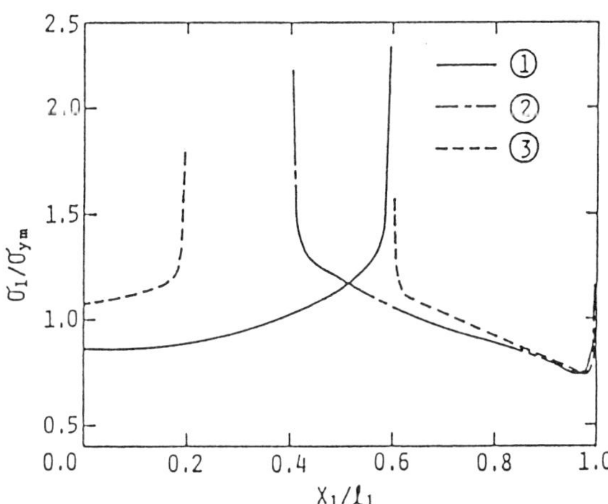

$$\left(\begin{array}{l} y_2 = h_2,\ h_1/h_2 = 25,\ E_1/E_2 = 10,\ h_1/l_1 = 1,\ \nu_1/\nu_2 = 1, \\ \text{On } |x_1| \leq l_1,\ F(x_1) \text{ is constant} \end{array}\right)$$

Fig.6 An effect of bonding regions and its position on maximum principal stress distribution at the interface

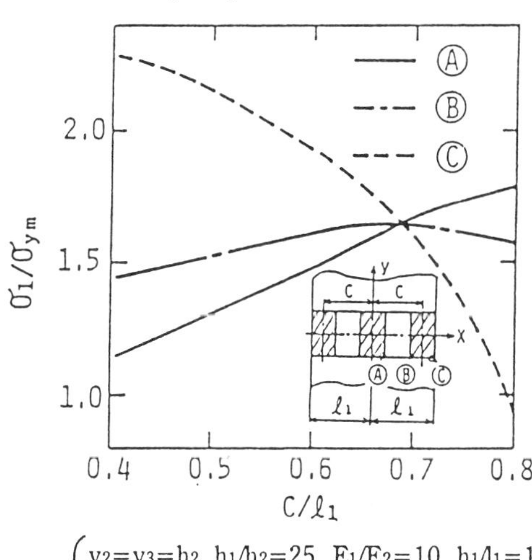

$$\left(\begin{array}{l} y_2 = y_3 = h_2,\ h_1/h_2 = 25,\ E_1/E_2 = 10,\ h_1/l_1 = 1, \\ \nu_1/\nu_2 = 1,\ l_2 = l_3 = 0.2l_1,\ \text{On } |x_1| \leq l_1, \\ F(x_1) \text{ is constant} \end{array}\right)$$

Fig.7 An effect of bonding position on maximum principal stress near the bond edges

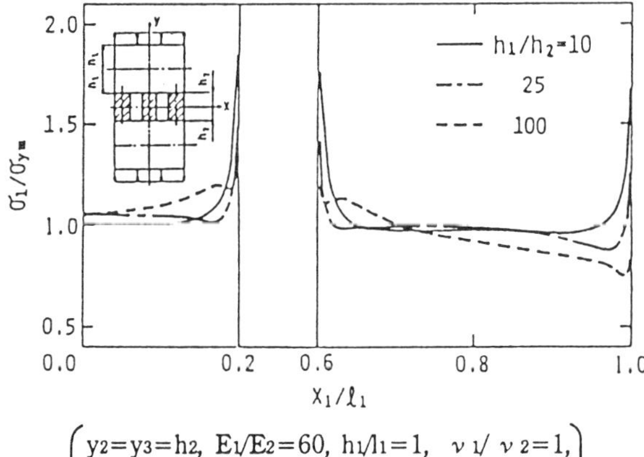

$$\left(\begin{array}{l} y_2 = y_3 = h_2,\ E_1/E_2 = 60,\ h_1/l_1 = 1,\ \nu_1/\nu_2 = 1, \\ l_2 = l_3 = 0.2l_1,\ C = 0.8l_1,\ \text{On } |x_1| \leq l_1,\ F(x_1) \\ \text{is constant} \end{array}\right)$$

Fig.5 An effect of thickness of adhesive on maximum principal stress distribution at the interface

29

0.2, 0.6 and 1.0, in logarithmic scales (Hattori,T. et al., 1988) in order to examine the property of the stress singularity at the close vicinity of a bonding edge, as is the case Fig. 4. Generally, the maximum principal stress singularity causes is expressed approximately by eq. (16).

$$\sigma_1(r) = K/r^\lambda \qquad (16)$$

where $\sigma_1(r)$: maximum principal stress, r: distance from singular point, K: intensity of stress singularity, λ : order of stress singularity.

The parameter λ is determined by the method which was demonstrated in the references (Timoshenko,S.P. and Goodier,J.N., 1970, Hattori,T. et al., 1988). In this paper,

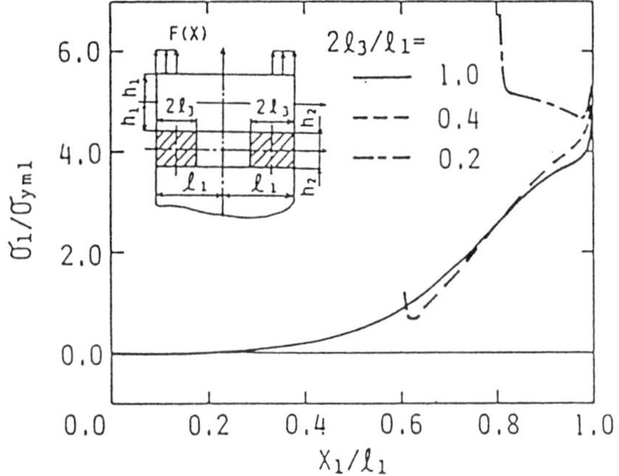

$$\begin{pmatrix} y_2=h_2,\ h_1/h_2=5,\ E_1/E_2=10,\ h_1/l_1=0.2,\ \nu_1/\nu_2=1, \\ l_2=l_3=0.2l_1,\ \text{On } 0.8l_1 \leq |x_1| \leq l_1,\ F(x_1) \text{ is constant} \end{pmatrix}$$

(a) in the case where bonding region is two

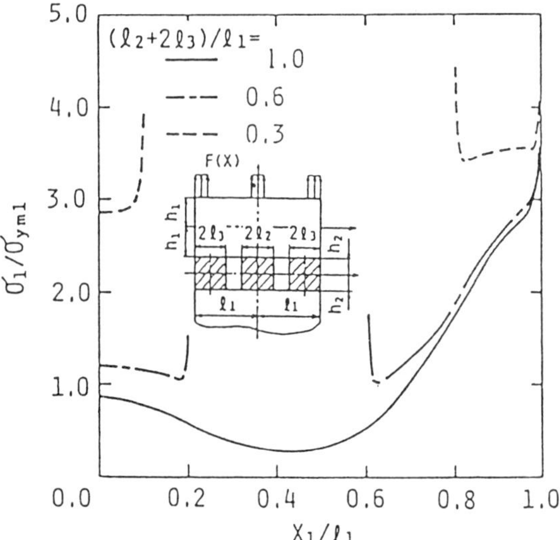

$$\begin{pmatrix} y_2=y_3=h_2,\ h_1/h_2=5,\ E_1/E_2=10,\ h_1/l_1=0.2, \\ \nu_1/\nu_2=1,\ \text{On } |x_1| \leq 0.05l_1,\ 0.9l_1 \leq |x_1| \leq l_1, \\ F(x_1) \text{ is constant} \end{pmatrix}$$

(b) in the case where bonding region is three

Fig.8 An effect of load distribution on maximum principal stress distribution at the the interface

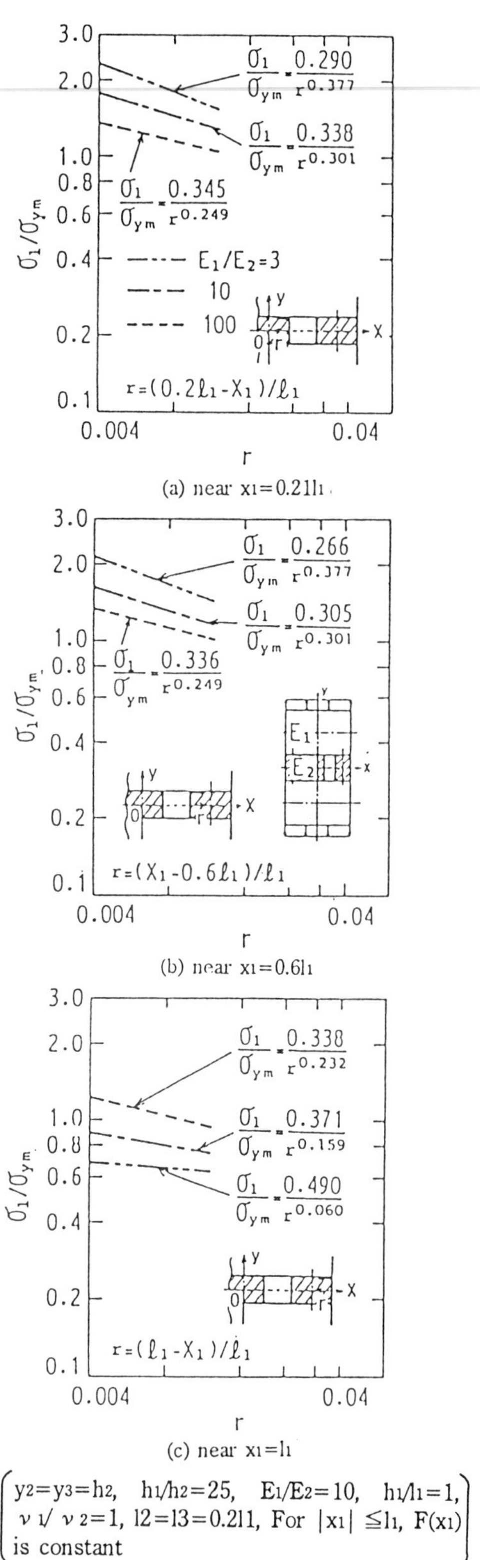

(a) near $x_1=0.21l_1$

(b) near $x_1=0.6l_1$

(c) near $x_1=l_1$

$$\begin{pmatrix} y_2=y_3=h_2,\ h_1/h_2=25,\ E_1/E_2=10,\ h_1/l_1=1, \\ \nu_1/\nu_2=1,\ l_2=l_3=0.2l_1,\ \text{For } |x_1| \leq l_1,\ F(x_1) \\ \text{is constant} \end{pmatrix}$$

Fig.9 Distribution of maximum principal stress near the edge of the interface indicated in logarithmic scales

r is varied between 0.004l_1 and 0.012l_1 as expressed by the equations r=(0.2l_1 $-$x$_1$)/l_1 , r=(x$_1$ $-$0.6l_1)/l_1 and r=(l_1 $-$x$_1$)/l_1 in cases where x$_1$ /l_1 is 0.2, 0.6 and 1.0, respectively. Determining the value of K from the distribution indicated in logarithmic scales, the equations of σ_1 are expressed in Fig. 9 for the value of E$_1$ /E$_2$ =3, 10 and 100.

4.2 Comparisons between the numerical and the experimental results

Figure 10(a) shows a comparison between the numerical and the experimental results with respect to the strains on the adherend in the y$_1$ direction where the position is 5mm away from the interface. The ordinate is the distance x$_1$ from the center of the specimen. A solid line represents the numerical results and the symbol o represents the experimental results. In the numerical calculations, Young's modulus and Poisson's ratio of the adherend and the adhesive are chosen as 206GPa, 0.3 and 3.6GPa, 0.38, respectively. The thickness 2h$_2$ of the adhesive was measured as 0.055mm. It is seen that the numerical results are satisfactorily consistent with the experimental ones in respect to the strains of the adherend.

Figure 10(b) shows the principal stress difference (σ_1 $-$ σ_2) at the interface (y$_2$ =y$_3$ =10mm). Fairly good agreement is seen between the numerical results and the experimental results in respect to the principal stress difference.

Table 1 shows the comparisons between the analytical and the experimental results concerning joint strength. Experiments on the joint strength were conducted 30 times for each test specimen. Statistical procedure was applied to the experimental results and a non ruptured value over 95% was determined as the joint strength. In the joint strength evaluation, fracture stress was previously measured by using thin hollow cylinder specimens (Sawa,T. and Kobayashi,T.). The fracture stress σ_* was measured as 47.4MPa for Al$-$Al specimen (Sawa,T. and Kobayashi,T.). In the analysis, joint strength is evaluated using the maximum principal stress at the interface. The joint strength is determined when the maximum principal stress at the interface of x$_1$ /l_1 =0.996 reaches the fracture stress σ_* . Fairly good agreement is seen between the analytical and the experimental results. It is shown that joint strength of the band adhesive joint is the same as that of the joint of which the interface is completely bonded.

5. CONCLUSIONS

In this paper, the stress distributions at the interface in band adhesive butt joints subjected to tensile loads were examined. The following results were obtained.

(1) By replacing adherends and an adhesive by finite strips, a method for analyzing the stress distributions at the interface was demonstrated using the two$-$dimensional theory of elasticity (plane stress).

(2) The effects of the ratio of Young's modulus for the adherends to that for the adhesive, the adhesive thickness, the adhesive region and it's position, and tensile load distributions on the stress distributions at the interface were shown by the numerical calculations. As a result, it is seen that the interface stress at the edge decrease with an increase of the ratio E$_1$ /E$_2$ and with a decrease of the adhesive thickness. In regard to a number of bonding regions, it is seen that the

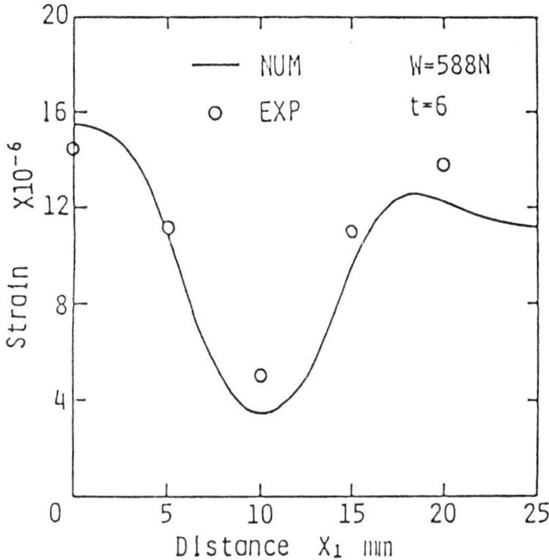

(a) the strains on adherend (y$_1$=$-$15mm)

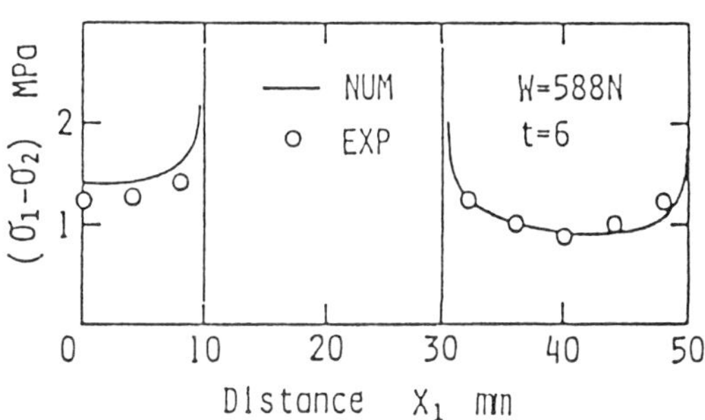

(b) a principal stress difference at the interface

Fig.10 Comparison of the numerical results with the experimental ones

Table 1 Comparison on joint strength (N)

		St$-$St	Al$-$Al
completely bonded (bonded area is 100%)	EXP	4371	4175
	NUM	4292 (11280)	4096 (11329)
partially bonded (bonded area is 60%)	EXP	4243	4096
	NUM	4185 (11162)	4047 (10408)

The values indicated in brackets are obtained from the theory of Mises

seen that the stress decrease at the edge of the interface with an increase of the numbers of bonding regions.

(3) It was discovered that the position of band bonding which made the value of the principal stress similar near the bond edges of the interface existed.

(4) It was seen that when the bonding position and its area were determined by taking into account the external load distributions, it might be possible that the band adhesive joint had the same strength as the case where the interface was bonded completely.

(5) Experiments concerning the strains on the adherend and photoelastic experiments were conducted. The analytical results were in fairy good agreement with the experimental results.

REFERENCES

Baker,R.M. and hatt,F.,1973,"Analysis of Bonded Joints in Vehicular Structures ",AIAA J.,Vol.11,No.12,p.1650.

Sen,J.K. and Jones,R.M,1980,"Stress in Double−Lap Joints Bonded with a Viscoelastic Adhesive : Part Ⅱ. Parametric Study and Joint Design ",AIAA J.,Vol.18,No.11,p.1376.

Adams,R.D. and Harris,J.A.,1987,"The Influence of Local Geometry on the Strength of Adhesive Joints",Int. J. Adhesion and Adhesives,Vol.7,No.2,pp.69−80.

Suzuki,Y.,1984,"Three−Dimensional Finite Element Analysis of Adhesive Scarf Joints of Steel Plates Loaded in Tension",Bull. of the JSME,Vol.27,No.231,p.1836.

Chow,C.L. and Woo,C.W.,1984,"Photoelastic Determination of Flaw Size and Distribution Effects on Adhesively Bonded Lap Joints",Int. Conger. Exp. Mech.,p.256.

Erdogan,F. and Ratwani,M.,1971,"Stress Distribution in Bonded Joints ",J.Composite Materials,5,p.378.

Wah,T.,1976,"The Adhesive Scarf Joint in Pure Bending",Int. J. Mech. Sci.,18,p.233.

Renton,W.J.,1979,"Analysis and Design of Adhesive Mecanical Chracterization Test Specimens",J. Adhesion,10,p.139.

Sawa,T. et al.,1987,"Stress Analysis of T−type Butt Adhesive Joint Subjected to External Bending Moments",J. Adhesion,24,p.1.

Sawa,T. et al.,1986,"Two−dimensional Stress Analysis of Adhesive Butt Joints Subjected to Tensile Loads",Bull. JSME,Vol.29,No.258,p.4037.

J.Pirvics,1974,"Two Dimensional Displacement−Stress Distributions in Adhesive Bonded Composite Structures",J. Adhesion,6,p.207.

Timoshenko,S.P. and Goodier,J.N.,1970,"Theory of Elasticiy (3rd ed)",McGraw−Hill,p.35.

Hattori,T. et al.,1988,"A Stress Singularity Parameter Approach for Evaluating Adhesive and Fretting Strength",Advances in Adhesively Bonded Joints.,The Winter Annnual Meeting of ASME,p.43.

Bogy,D.B.,1971,"Two Edge−Bonded Elastic Wedges of Different Materials and Wedge Angles Under Surface Tractions",J. Appl. Mech.,Vol.38,p.377.

Sawa,T. and Kobayashi,T. ,1988,"The Strength of Joints Combining an Adhesive with a Bolt", J.Adhesion,Vol.25,pp.269−280.

APPENDIX

The relationship among the unknown coefficients shown in Eqs.(6)−(15) are reduced as follows:

$$\bar{A}_n^{I} + \sum_{s=1} \bar{B}_s^{I} \bar{P}_{ns}^{I} = 0, \quad \bar{A}_n^{I} - \sum_{s=1} \bar{B}_s^{I} \bar{P}_{ns}^{I} = 0, \quad \bar{A}_n^{I} + \sum_{s=1} \bar{B}_s^{I} \bar{S}_{ns}^{I} = 0, \quad \dot{A}_n^{I} + \sum_{s=1} \dot{B}_s^{I} \dot{S}_{ns}^{I} = 0$$

$$\bar{A}_n^{II} + \sum_{s=1} \bar{B}_s^{II} \bar{P}_{ns}^{II} = 0, \quad \bar{A}_n^{II} + \sum_{s=1} \bar{B}_s^{II} \bar{S}_{ns}^{II} = 0, \quad \bar{A}_n^{III} + \sum_{s=1} \bar{B}_s^{III} \bar{P}_{ns}^{III} = 0, \quad \bar{A}_n^{III'} - \sum_{s=1} \bar{B}_s^{III'} \bar{P}_{ns}^{III'} = 0$$

$$\bar{A}_n^{III} + \sum_{s=1} \bar{B}_s^{III} \bar{S}_{ns}^{III} = 0, \quad \bar{A}_n^{III'} + \sum_{s=1} \bar{B}_s^{III'} S_{ns}^{III'} = 0, \quad \sum_{n=1} A_n^{I} \bar{Q}_{sn}^{I} + \bar{B}_s^{I} - \sum_{n=1} \bar{A}_n^{I} \bar{Q}_{sn}^{I} + \bar{B}_s^{I} = - a_s$$

$$\sum_{n=1} \bar{A}_n^{I} \tilde{R}_{sn}^{I} + \tilde{B}_s^{I} + \sum_{n=1} \dot{A}_n^{I} \dot{R}_{sn}^{I} + \dot{B}_s^{I} = 0$$

$$- \sum_{n=1} \bar{A}_n^{I} \bar{Q}_{sn}^{I} - \bar{B}_s^{I} - \sum_{n=1} \bar{A}_n^{I} \bar{Q}_{sn}^{I} + \bar{B}_s^{I} + \sum_{m=1} \left[\sum_{n=1} \bar{A}_n^{II} Q_{mn}^{II} + \bar{B}_m^{II} \right] \bar{a}_{sm}^{II} + 2 \sum_{m=1} \left[\sum_{n=1} \bar{A}_n^{III} Q_{mn}^{III} + \bar{B}_m^{III} \right] \bar{a}_{sm}^{III}$$

$$- 2 \sum_{m=1} \left[\sum_{n=1} \bar{A}_n^{III'} \bar{Q}_{mn}^{III'} - \bar{B}_m^{III'} \right] \bar{a}_{sm}^{III'} = b_s + 2 C_s$$

$$- \sum_{n=1} \bar{A}_n^{I} \dot{R}_{sn}^{I} - \tilde{B}_s^{I} + \sum_{n=1} \dot{A}_n^{I} \dot{R}_{sn}^{I} + \dot{B}_s^{I} - \sum_{m=1} \left[\sum_{n=1} \bar{A}_n^{II} \bar{R}_{mn}^{II} + \bar{B}_m^{II} \right] \bar{a}_{sm}^{II} - 2 \sum_{m=1} \left[\sum_{n=1} \bar{A}_n^{III} \bar{R}_{mn}^{III} + \bar{B}_m^{III} \right] \bar{a}_{sm}^{III}$$

$$+ 2 \sum_{m=1} \left[\sum_{n=1} \bar{A}_n^{III'} \bar{R}_{mn}^{III'} + \bar{B}_s^{III'} \right] \dot{a}_{sm}^{III'} = 0$$

$$\sum_{m=1} K_m a_{sm}^{I} + \sum_{m=1} L_m a_{sm}^{I} - \sum_{n=1} \bar{A}_n^{II} U_{sn}^{II} - \sum_{m=1} \bar{B}_m^{II} V_m^{II} \bar{b}_{sm}^{II} - \bar{B}_s^{II} \bar{W}_s^{II} = d s$$

$$\sum_{m=1} M_m b_{sm}^{I} - \sum_{m=1} N_m b_{sm}^{I} - \bar{B}_s^{II} \bar{F}_s^{II} - \sum_{n=1} \bar{A}_n^{II} \bar{H}_{sn}^{II} - \sum_{m=1} \bar{B}_m^{II} \bar{J}_m^{II} \bar{b}_{sm}^{II} = 0$$

$$\sum_{m=1} K_m C_{sm}^{I} + \sum_{m=1} L_m C_{sm}^{I} - \sum_{n=1} \bar{A}_n^{III} U_{sn}^{III} - \sum_{m=1} \bar{B}_m^{III} V_m^{III} \bar{b}_{sm}^{III} - \bar{B}_s^{III} \bar{W}_s^{III} = e_s$$

$$\sum_{m=1} K_m \bar{C}_{sm}^{I} + \sum_{m=1} L_m \bar{C}_{sm}^{I} + \sum_{n=1} \bar{A}_n^{III'} U_{sn}^{III'} + \sum_{m=1} \bar{B}_m^{III'} V_m^{III'} \bar{b}_{sm}^{III'} + \bar{B}_s^{III'} \bar{W}_s^{III'} = 0$$

$$\sum_{m=1} M_m d_{sm}^{I} - \sum_{m=1} N_m d_{sm}^{I} - \bar{B}_s^{III} \bar{F}_s^{III} - \sum_{n=1} \bar{A}_n^{III} \bar{H}_{sn}^{III} - \sum_{m=1} \bar{B}_m^{III} \bar{J}_m^{III} \bar{b}_{sm}^{III} = 0$$

$$\sum_{m=1} M_m \bar{d}_{sm}^{I} - \sum_{m=1} N_m \bar{d}_{sm}^{I} - \bar{B}_s^{III'} \bar{F}_s^{III'} + \sum_{n=1} \bar{A}_n^{III'} \bar{H}_{sn}^{III'} + \sum_{m=1} + \sum_{m=1} \bar{B}_m^{III'} \bar{J}_m^{III'} \bar{b}_{sm}^{III'} = 0 \tag{17}$$

$$A_0^{I} = a_0, \quad A_0^{I} = \frac{l_2}{l_1} A_0^{III} + \frac{2l_3}{l_1} A_0^{III} \tag{18}$$

where

$$K_s = \sum_{n=1} \bar{A}_n U_{sn}^{I} + \sum_{n=1} \bar{A}_n U_{sn}^{I} + \bar{B}_s^{I} \bar{W}_s^{I} - \dot{B}_s^{I} \dot{W}_s, \quad L_s = \bar{B}_s^{I} V_s^{I} + \bar{B}_s^{I} V_s^{I}$$

$$M_s = - \bar{B}_s^{I} F_s^{I} + \bar{B}_s^{I} F_s^{I} - \sum_{n=1} \bar{A}_n^{I} \bar{H}_{sn}^{I} + \sum_{n=1} \dot{A}_n^{I} \dot{H}_{sn}^{I}, \quad N_s = \bar{B}_s^{I} J_s^{I} + \dot{B}_s^{I} \dot{J}_s^{I}$$

$$\bar{P}_{ns}^{I} = P_{ns}(a_n^{I}, \lambda_s^{I}, h_1, \bar{\Omega}_s^{I}) = \frac{4(-1)^{n+1} a_n^{I2} \lambda_s^{I} \, \text{sh}(\lambda_s^{I} h_1)}{\bar{\Omega}_s^{I}(a_n^{I2} + \lambda_s^{I2})^2 h_1}, \quad \bar{P}_{ns}^{II} = P_{ns}(a_n^{II}, \lambda_s^{II}, h_2, \bar{\Omega}_s^{II})$$

$$\bar{P}_{ns}^{III} = P_{ns}(a_n^{III}, \lambda_s^{III}, h_2, \bar{\Omega}_s^{III}), \quad \bar{P}_{ns}^{III'} = P_{ns}(a_n^{III}, \lambda_s^{III'}, h_2, \bar{\Omega}_s^{III'}), \quad \dot{S}_{ns}^{I} = P_{ns}(a_n^{I}, \lambda_s^{I'}, h_1, \dot{\Omega}_s^{I})$$

$$\dot{S}_{ns}^{I} = S_{ns}(a_n^{I'}, \lambda_s^{I'}, h_1, \bar{\Omega}_s^{I}) = \frac{4(-1)^{n+s} a_n^{I2} \lambda_s^{I'} \, \text{ch}^2(\lambda_s^{I'} h_1)}{\bar{\Omega}_s^{I}(a_n^{I'2} + \lambda_s^{I'2}) h_1}, \quad \bar{S}_{ns}^{II} = S_{ns}(a_n^{II'}, \lambda_s^{II'}, h_2, \bar{\Omega}_s^{II})$$

$$\bar{S}_{ns}^{III} = S_{ns}(a_n^{III'}, \lambda_s^{III'}, h_2, \bar{\Omega}_s^{III'}), \quad \bar{S}_{ns}^{III'} = S_{ns}(a_n^{III'}, \lambda_s^{III'}, h_2, \bar{\Omega}_s^{III'}), \quad \bar{P}_{ns}^{I} = S_{ns}(a_n^{I'}, \lambda_s^{I}, h_1, \bar{\Omega}_s^{I})$$

$$\bar{Q}_{sn}^{I} = Q_{sn}(a_n^{I}, \lambda_s^{I}, l_1, \bar{\Delta}_n) = \frac{4(-1)^{n+s} a_n^{I} \lambda_s^{I2} \, \text{sh}^2(a_n^{I} l_1)}{\bar{\Delta}_s(a_n^{I2} + \lambda_s^{I2})^2 l_1}, \quad \bar{Q}_{sn}^{I} = Q_{sn}(a_n^{I'}, \lambda_s^{I}, l_1, \bar{\Delta}_n)$$

$$\bar{Q}_{sn}^{II} = Q_{sn}(a_n^{II}, \lambda_s^{II}, l_2, \bar{\Delta}_n^{II}), \quad Q_{sn}^{III} = Q_{sn}(a_n^{III}, \lambda_s^{III}, l_3, \bar{\Delta}_n^{III}), \quad \bar{R}_{sn}^{III'} = Q_{sn}(a_n^{III'}, \lambda_s^{III'}, l_3, \bar{\Delta}_n^{III'})$$

$$\bar{R}_{sn}^{I} = R_{sn}(a_n^{I'}, \lambda_s^{I'}, l_1, \bar{\Delta}_n^{I}) = \frac{4(-1)^{n+s} a_n^{I'} \lambda_s^{I'2} \, \text{ch}^2(a_n^{I'} l_1)}{\bar{\Delta}_n^{I}(a_n^{I'2} + \lambda_s^{I'2})^2 l_1}, \quad \dot{R}_{sn}^{I} = R_{sn}(a_n^{I}, \lambda_s^{I'}, l_1, \dot{\Delta}_n^{I})$$

$$\bar{R}_{sn}^{II} = R_{sn}(a_n^{II'}, \lambda_s^{II'}, l_2, \bar{\Delta}_n^{II}), \quad \bar{R}_{sn}^{III} = R_{sn}(a_n^{III'}, \lambda_s^{III'}, l_3, \bar{\Delta}_n^{III}), \quad \bar{Q}_{sn}^{III'} = R_{sn}(a_n^{III'}, \lambda_s^{III'}, l_3, \bar{\Delta}_n^{III'})$$

$$U_{sn}^{I} = U_{sn}(a_n^{I}, \lambda_s^{I'}, l_1, \bar{\Delta}_n, G_1, \nu_1)$$

$$= \frac{2(-1)^{s+n}}{\bar{\Delta}_n^{I}(a_n^{I2} + \lambda_s^{I'2}) G_1 l_1} \left\{ a_n^{I} l_1 + \left(\frac{2}{1+\nu_1} - \frac{\lambda_s^{I'2} - a_n^{I2}}{a_n^{I2} + \lambda_s^{I'2}} \right) \text{sh}(a_n^{I} l_1) \, \text{ch}(a_n^{I} l_1) \right\}$$

$$U_{sn}^{I} = U_{sn}(a_n^{I'}, \lambda_s^{I'}, l_1, \bar{\Delta}_n, G_1, \nu_1), \quad U_{sn}^{II} = U_{sn}(a_n^{II}, \lambda_s^{II}, l_1, \bar{\Delta}_n^{I}, G_2, \nu_2)$$

$$U_{sn}^{III} = U_{sn}(a_n^{III}, \lambda_s^{III}, l_3, \bar{\Delta}_n^{II}, G_2, \nu_2)$$

$$U_{sn}^{III'} = U_{sn}(a_n^{III'}, \lambda_s^{III'}, l_3, \bar{\Delta}_n^{II}, G_2, \nu_2)$$

$$= \frac{2(-1)^{s+n}}{\bar{\Delta}_n^{I}(a_n^{III'2} + \lambda_s^{III'2}) G_2 l_3} \left\{ \left(\frac{2}{1+\nu_2} - \frac{\lambda_s^{III'2} - a_n^{III'2}}{a_n^{III'2} + \lambda_s^{III'2}} \right) \text{ch}(a_n^{III'} l_3) \, \text{sh}(a_n^{III'} l_3) - a_n^{III'} l_3 \right\}$$

$$\bar{W}_s^{I} = W_s(\lambda_s^{I}, h_1, \bar{\Omega}_s^{I}, G_1, \nu_1) = \frac{2 \, \text{ch}^2(\lambda_s^{I} h_1)}{\bar{\Omega}_s^{I} \lambda_s^{I} G_1(1+\nu_1)}, \quad \dot{W}_s^{I} = W_s(\lambda_s^{I}, h_1, \dot{\Omega}_s^{I}, G_1, \nu_1) = \frac{2 \, \text{sh}^2(\lambda_s^{I} h_1)}{\dot{\Omega}_s^{I} \lambda_s^{I} G_1(1+\nu_1)}$$

$$\bar{W}_s^{II} = W_s(\lambda_s^{II}, h_2, \bar{\Omega}_s^{II}, G_2, \nu_2), \quad \bar{W}_s^{III} = W_s(\lambda_s^{III}, h_2, \bar{\Omega}_s^{III}, G_2, \nu_2), \quad \bar{W}_s^{III'} = W_s(\lambda_s^{III'}, h_2, \bar{\Omega}_s^{III'}, G_2, \nu_2)$$

$$V_s^{I} = V_s(\lambda_s^{I}, h_1, \bar{\Omega}_s^{I}, G_1, \nu_1) = \frac{1}{\bar{\Omega}_s^{I} \lambda_s^{I} G_1} \left\{ \lambda_s^{I} h_1 + \frac{\nu_1 - 1}{1+\nu_1} \, \text{sh}(\lambda_s^{I} h_1) \, \text{ch}(\lambda_s^{I} h_1) \right\}$$

$$\bar{V}_s^{\mathrm{I}} = V_s(\lambda_s^{\mathrm{I}}, h_1, \bar{\Omega}_s^{\mathrm{I}}, G_1, \nu_1) = \frac{1}{\bar{\Omega}_s^{\mathrm{I}} \lambda_s^{\mathrm{I}} G_1} \left\{ \lambda_s^{\mathrm{I}} h_1 - \frac{\nu_1 - 1}{1 + \nu_1} \operatorname{sh}(\lambda_s^{\mathrm{I}} h_1) \operatorname{ch}(\lambda_s^{\mathrm{I}} h_1) \right\}$$

$$\bar{V}_s^{\mathrm{II}} = V_s(\lambda_s^{\mathrm{II}}, h_2, \bar{\Omega}_s^{\mathrm{II}}, G_2, \nu_2), \quad \bar{v}_s^{\mathrm{III}} = V_s(\lambda_s^{\mathrm{III}}, h_2, \bar{\Omega}_s^{\mathrm{III}}, G_2, \nu_2), \quad \bar{V}_s^{\mathrm{III'}} = V_s(\lambda_s^{\mathrm{III'}}, h_2, \bar{\Omega}_s^{\mathrm{III'}}, G_2, \nu_2)$$

$$\bar{H}_{sn}^{\mathrm{I}} = H_{sn}(a_n^{\mathrm{I'}}, \lambda_s^{\mathrm{I}}, l, \bar{\Delta}_n^{\mathrm{I}}, G_1, \nu_1) = \frac{2(-1)^{s+n}\lambda_s^{\mathrm{I}}}{\bar{\Delta}_n^{\mathrm{I}}(a_n^{\mathrm{I'2}} + \lambda_s^{\mathrm{I'2}}) G_1 l_1} \left\{ a_n^{\mathrm{I'}} l_1 + \left\{ \frac{\nu + 3}{1 + \nu_1} - \frac{2 a_n^{\mathrm{I'2}}}{a_n^{\mathrm{I'2}} + \lambda_s^{\mathrm{I'2}}} \right\} \operatorname{sh}(a_n^{\mathrm{I'}} l_1) \operatorname{ch}(a_n^{\mathrm{I'}} l_1)$$

$$\tilde{H}_{sn}^{\mathrm{I}} = H_{sn}(a_n^{\mathrm{I'}}, \lambda_s^{\mathrm{I}}, l_1, \tilde{\Delta}_n^{\mathrm{I}}, G_1, \nu_1), \quad \bar{H}_{sh}^{\mathrm{II}} = H_{sn}(a_n^{\mathrm{II'}}, \lambda_s^{\mathrm{II}}, l_2, \tilde{\Delta}_n^{\mathrm{II}}, G_2, \nu_2)$$

$$\bar{H}_{sn}^{\mathrm{III}} = H_{sn}(a_n^{\mathrm{III}}, \lambda_s^{\mathrm{III}}, l_3, \tilde{\Delta}_n^{\mathrm{III}}, G_2, \nu_2)$$

$$\bar{H}_{sn}^{\mathrm{III'}} = H_{sn}(a_n^{\mathrm{III'}}, \lambda_s^{\mathrm{III'}}, l_3, \tilde{\Delta}_n^{\mathrm{III'}}, G_2, \nu_2)$$

$$= \frac{2(-1)^{s+n}\lambda_s^{\mathrm{III'}}}{\tilde{\Delta}_n^{\mathrm{III'}}(a_n^{\mathrm{III'2}} + \lambda_s^{\mathrm{III'2}}) G_2 l_3} \left\{ a_n^{\mathrm{III'}} l_3 - \left(\frac{2 \nu_2}{1 + \nu_2} - 2 \frac{a_n^{\mathrm{III'2}}}{a_n^{\mathrm{III'2}} + \lambda_s^{\mathrm{III'2}}} \right) \operatorname{ch}(a_n^{\mathrm{III'}} l_3) \operatorname{sh}(a_n^{\mathrm{III'}} l_3) \right\}$$

$$F_s^{\mathrm{I}} = \lambda_s^{\mathrm{I}} \times W_s(\lambda_s^{\mathrm{I}}, h_1, \bar{\Omega}_s^{\mathrm{I}}, G_1, \nu_1), \quad \bar{F}_s^{\mathrm{I}} = \lambda_s^{\mathrm{I}} \times W_s(\lambda_s^{\mathrm{I}}, h_1, \bar{\Omega}_s^{\mathrm{I}}, G_1, \nu_1)$$

$$F_s^{\mathrm{II}} = \lambda_s^{\mathrm{II}} \times W_s(\lambda_s^{\mathrm{II}}, h_2, \bar{\Omega}_s^{\mathrm{II}}, G_2, \nu_2), \quad \bar{F}_s^{\mathrm{III}} = \lambda_s^{\mathrm{III}} \times W_s(\lambda_s^{\mathrm{III}}, h_2, \bar{\Omega}_s^{\mathrm{III}}, G_2, \nu_2)$$

$$F_s^{\mathrm{III'}} = \lambda_s^{\mathrm{III'}} \times W_s(\lambda_s^{\mathrm{III'}}, h_2, \bar{\Omega}_s^{\mathrm{III'}}, G_2, \nu_2), \quad \bar{J}_s^{\mathrm{I}} = \lambda_s^{\mathrm{I'}} \times V_s(\lambda_s^{\mathrm{I'}}, h_1, \tilde{\Omega}_s^{\mathrm{I}}, G_1, \nu_1)$$

$$\tilde{J}_s^{\mathrm{I}} = \lambda_s^{\mathrm{I'}} \times V_s(\lambda_s^{\mathrm{I'}}, h_1, \tilde{\Omega}_s^{\mathrm{I}}, G_1, \nu_1), \quad J_s^{\mathrm{II}} = \lambda_s^{\mathrm{II'}} \times V_s(\lambda_s^{\mathrm{II'}}, h_2, \tilde{\Omega}_s^{\mathrm{II}}, G_2, \nu_2)$$

$$\tilde{J}_s^{\mathrm{III}} = \lambda_s^{\mathrm{III'}} \times V_s(\lambda_s^{\mathrm{III'}}, h_2, \tilde{\Omega}_s^{\mathrm{III}}, G_2, \nu_2), \quad \tilde{J}_s^{\mathrm{III'}} = \lambda_s^{\mathrm{III'}} \times V_s(\lambda_s^{\mathrm{III'}}, h_2, \tilde{\Omega}_s^{\mathrm{III'}} G_2, \nu_2)$$

$$\bar{a}_{sn}^{\mathrm{II}} = a_{sn}(\lambda_s^{\mathrm{I}}, \lambda_m^{\mathrm{II}}, l_1, l_2) = \frac{\sin[(\lambda_s^{\mathrm{I}} + \lambda_m^{\mathrm{II}}) l_2]}{(\lambda_s^{\mathrm{I}} + \lambda_m^{\mathrm{II}}) l_1} + \frac{\sin[(\lambda_s^{\mathrm{I}} - \lambda_m^{\mathrm{II}}) l_2]}{(\lambda_s^{\mathrm{I}} - \lambda_m^{\mathrm{II}}) l_1}$$

$$\bar{a}_{sn}^{\mathrm{II}} = b_{sn}(\lambda_s^{\mathrm{I'}}, \lambda_m^{\mathrm{II'}}, l_1, l_2) = \frac{\sin[(\lambda_s^{\mathrm{I'}} - \lambda_m^{\mathrm{II'}}) l_2]}{(\lambda_s^{\mathrm{I'}} - \lambda_m^{\mathrm{II'}}) l_1} - \frac{\sin[(\lambda_s^{\mathrm{I'}} + \lambda_m^{\mathrm{II'}}) l_2]}{(\lambda_s^{\mathrm{I'}} + \lambda_m^{\mathrm{II'}}) l_1}$$

$$\bar{a}_{sn}^{\mathrm{III}} = \cos(\lambda_s^{\mathrm{I}} c) \times a_{sn}(\lambda_s^{\mathrm{I}}, \lambda_m^{\mathrm{III}}, l_1, l_3), \quad \bar{a}_{sn}^{\mathrm{III'}} = -\sin(\lambda_s^{\mathrm{I}} c) \times b_{sn}(\lambda_s^{\mathrm{I}}, \lambda_m^{\mathrm{III'}}, l_1, l_3)$$

$$\bar{a}_{sn}^{\mathrm{III}} = \cos(\lambda_s^{\mathrm{I'}} c) \times b_{sn}(\lambda_s^{\mathrm{I'}}, \lambda_m^{\mathrm{III'}}, l_1, l_3), \quad \bar{a}_{sn}^{\mathrm{III'}} = \sin(\lambda_s^{\mathrm{I}} c) \times a_{sn}(\lambda_s^{\mathrm{I'}}, \lambda_m^{\mathrm{III}}, l_1, l_3)$$

$$\bar{b}_{sn}^{\mathrm{II}} = b_{sn}(\lambda_s^{\mathrm{II'}}, \lambda_m^{\mathrm{II}}, l_2, l_2), \quad \bar{b}_{sn}^{\mathrm{II}} = b_{sn}(\lambda_s^{\mathrm{II}}, \lambda_m^{\mathrm{II'}}, l_2, l_2), \quad \bar{b}_{sn}^{\mathrm{III}} = b_{sn}(\lambda_s^{\mathrm{III'}}, \lambda_m^{\mathrm{III}}, l_3, l_3)$$

$$\bar{b}_{sn}^{\mathrm{III}} = a_{sn}(\lambda_s^{\mathrm{III}}, \lambda_m^{\mathrm{III'}}, l_3, l_3), \quad \bar{b}_{sn}^{\mathrm{III}} = b_{sn}(\lambda_s^{\mathrm{III}}, \lambda_m^{\mathrm{III'}}, l_3, l_3), \quad \bar{b}_{sn}^{\mathrm{III'}} = a_{sn}(\lambda_s^{\mathrm{III'}}, \lambda_m^{\mathrm{III}}, l_3, l_3)$$

$$a_{sn}^{\mathrm{I}} = b_{sn}(\lambda_s^{\mathrm{II'}}, \lambda_m^{\mathrm{I}}, l_2, l_2), \quad a_{sn}^{\mathrm{I'}} = b_{sn}(\lambda_s^{\mathrm{II'}}, \lambda_m^{\mathrm{I'}}, l_2, l_2), \quad b_{sn}^{\mathrm{I}} = b_{sn}(\lambda_s^{\mathrm{II}}, \lambda_m^{\mathrm{I'}}, l_2, l_2)$$

$$b_{sn}^{\mathrm{I'}} = b_{sn}(\lambda_s^{\mathrm{II}}, \lambda_m^{\mathrm{I}}, l_2, l_2), \quad c_{sn}^{\mathrm{I}} = \cos(\lambda_m^{\mathrm{I}} c) \times b_{sn}(\lambda_s^{\mathrm{III'}}, \lambda_m^{\mathrm{I}}, l_3, l_3)$$

$$c_{sn}^{\mathrm{I'}} = \cos(\lambda_m^{\mathrm{I'}} c) \times b_{sn}(\lambda_s^{\mathrm{III'}}, \lambda_m^{\mathrm{I'}}, l_3, l_3), \quad \bar{c}_{sn}^{\mathrm{I}} = \sin(\lambda_m^{\mathrm{I}} c) \times a_{sn}(\lambda_s^{\mathrm{III}}, \lambda_m^{\mathrm{I}}, l_3, l_3)$$

$$\bar{c}_{sn}^{\mathrm{I'}} = \sin(\lambda_m^{\mathrm{I'}} c) \times b_{sn}(\lambda_s^{\mathrm{III}}, \lambda_m^{\mathrm{I'}}, l_3, l_3), \quad d_{sn}^{\mathrm{I}} = \cos(\lambda_m^{\mathrm{I}} c) \times b_{sn}(\lambda_s^{\mathrm{III}}, \lambda_m^{\mathrm{I}}, l_3, l_3)$$

$$d_{sn}^{\mathrm{I'}} = \cos(\lambda_m^{\mathrm{I}} c) \times b_{sn}(\lambda_s^{\mathrm{III}}, \lambda_m^{\mathrm{I}}, l_3, l_3), \quad \bar{d}_{sn}^{\mathrm{I}} = \sin(\lambda_m^{\mathrm{I'}} c) \times a_{sn}(\lambda_s^{\mathrm{III'}}, \lambda_m^{\mathrm{I'}}, l_3, l_3)$$

$$\bar{d}_{sn}^{\mathrm{III'}} = \sin(\lambda_m^{\mathrm{I}} c) \times a_{sn}(\lambda_s^{\mathrm{III'}}, \lambda_m^{\mathrm{I}}, l_3, l_3)$$

$$b_s = \frac{2 A_0^{\mathrm{II}}}{\lambda_s^{\mathrm{I}} l_1} \sin(\lambda_s^{\mathrm{I}} l_2), \quad c_s = \frac{2 A_0^{\mathrm{III}}}{\lambda_s^{\mathrm{I}} l_1} \cos(\lambda_s^{\mathrm{I}} a) \sin(\lambda_s^{\mathrm{I}} l_3)$$

$$d_s = \frac{4(-1)^s}{\lambda_s^{\mathrm{II'2}} l_2} \left(\frac{\nu^2 A_0^{\mathrm{II}}}{E_2} - \frac{\nu_1 A_0^{\mathrm{I}}}{E_1} \right), \quad e_s = \frac{4(-1)^s}{\lambda_s^{\mathrm{III'2}} l_3} \left(\frac{\nu^2 A_0^{\mathrm{III}}}{E_2} - \frac{\nu_1 A_0^{\mathrm{I}}}{E_1} \right), \quad G_1 = \frac{E_1}{2(1 + \nu_1)}, \quad G_2 = \frac{E_2}{2(1 + \nu_2)}$$

DE-Vol. 87, Reliability, Stress Analysis, and Failure Prevention
Issues in Emerging Technologies and Materials
ASME 1995

Elastoplastic Finite Element Analysis and Strength Evaluation of Adhesive Butt Joints of Dissimilar Hollow Shafts Subjected to External Bending Moments

Toshiyuki SAWA
Department of Mechanical Engineering
Yamanashi University
Takeda, Kofu, Yamanashi
Japan

Mitsuhiro AOKI
Department of Mechanical Engineering
Yamanashi University
Takeda, Kofu, Yamanashi
Japan

Osamu NISHIKAWA
Department of Mechanical Engineering
Yamanashi University
Takeda, Kofu, Yamanashi
Japan

Abstract

Stress distributions and deformation of adhesive butt joints are analyzed by elastoplastic finite element method when the joints of dissimilar shafts are subjected to external bending moments. The effects of ratio of Young's moduli for adherends to that for an adhesive and the adhesive thickness on the interface stress distribution are investigated. Joint strength is predicted by using the elastoplastic interface stress distributions. As the results, it is found that the singular stress at the edge of the interfaces increases with an increase of the ratio of Young's moduli. Measurement of strains in joints and experiments on joint strength were conducted. The numerical results are in fairly good agreement with the experimental results. It is observed that the joint strength of dissimilar shafts are smaller than that of similar shafts. A fracture of dissimilar adhesive shafts occurred from the interface of the adherend with smaller Young's modulus. It is seen that joint strength increases as the adhesive thickness increases.

1. Introduction

Recently, adhesive joints have been used in mechanical structures as the performance of bonding materials has advanced. However, the design of the joints results almost entirely from experience. It is expected to make data available for design and to establish an optimal design method. In establishing the design method for adhesive joints, it is necessary to know the stress distributions in joints more precisely. Many investigations have been carried out on lap, scarf, and butt adhesive joints subjected to tensile, bending moment, cleavage, and torsional loads (Wah, 1973, Adams and Peppiatt, 1974, Tsai and Morton, 1994, Sawa et al., 1987 and

Temma et al., 1990). But these investigation have treated two–dimensional configuration joints and three–dimensional configuration joints subjected to axi–symmetric loads (Sawa et al., 1990, Sawa et al., 1989, Temma et al., 1994, Nakano et al., 1991), and few have been carried out on the mechanical behaviors of three–dimensional configuration joints subjected to axi–asymmetric loads such as bending moments.

In this paper, stress distributions of adhesive butt joints consisted of dissimilar hollow shafts are analyzed by elastoplastic finite element method when the joints are subjected to external bending moments. The effects of the ratio of Young's moduli for adherends to that for an adhesive and the adhesive thickness on the interface stress distributions are investigated. Changes in yielded area at the interface are also shown with an increase of bending moment. Using the interface stress distributions, joint strength is estimated. For verification, measurement of strain in joints was conducted. Four–point bending tests were carried out to measure the joint strength. The analytical results are compared with the experimental ones.

2. Method for Analysis

Figure 1 shows an adhesive butt joint, consisted of two dissimilar finite hollow cylinders, subjected to an external bending moment. In order to know the stress distribution in adhesive [II], the elastoplastic finite element analysis is carried out. The adherends are denoted as finite hollow cylinders [I] and [III], and the adhesive as finite hollow cylinder [II]. The inner diameter of cylinder [I] is designated as $2a$ and the outer diameter as $2b$, the height as h_1, Young's modulus as E_1 and Poisson's ratio as ν_1. Those for cylinder [II] are designated as $2a$, $2b$, h_2, E_2 and ν_2, and those for cylinder [III] as $2a$,

2b, h_3, E_3 and ν_3, respectively. Figure 2 shows an example of mesh division (a half part of an adhesive joint) in FEM analysis. Hexahedron elements are employed. Finite elements and nodes employed are 2660 and 3432, respectively. When the inner diameter 2a is zero, that is, the solid shaft, pentahedron and hexahedron elements are used. Numbers of elements and nodes are 2660 and 3042, respectively. The FEM code employed is KSWAD/FEM·SOLVE(KUBOTA, Co., Ltd.).

An external bending moment M is assumed to be applied with liner stress distribution σ_z as shown in Fig. 1.

It is assumed that the stress σ_z acts on both ends of cylinders [I] and [III] within the region $a \leq r \leq b$ as shown in Figs. 1 and 2. The absolute value at the position r=b and θ =3/2· π is denoted as σ_{zmax} as shown in Figs. 1 and 2. A relationship between σ_z and M is expressed by Eq. (1).

$$M = 4 \int_0^{\frac{\pi}{2}} \int_a^b \sigma_z r^2 \cos\theta \, dr \, d\theta \qquad (1)$$

where

$$\sigma_z = -\frac{\sigma_{zmax}}{b} \cos\theta \times r \qquad (2)$$

In the numerical calculations, the dimensions of adhesive joints are chosen as 2a=20[mm], 2b=30[mm], h_1 =50[mm], h_2 =0.05[mm] and h_3 =50[mm]. The combinations of materials for adherends are steel and steel, aluminum and aluminum, and steel and aluminum, respectively. Epoxy bonding, which is solid at room temperature, was used. These combinations for adherend materials are denoted by St−St, Al−Al and St−Al. Table 1 shows the mechanical properties of the adherends and the adhesive. The stress−strain diagram is approximated as bilinear elastic−plastic body. The slope of the stress−strain diagram after yielding is denoted by c and the value is shown in Table 1. Figure 3 shows a sample stress−strain diagram for epoxy resin bonding. The dotted line indicates the approximated bilinear elastic−plastic behavior. The yield stress and the fracture stress of epoxy resin bond were obtained from experiments as 21.2[MPa] and 28.1[MPa], respectively.

3. Experimental Method

Experiments were. conducted to measure strain at the outer surface of the adherend and joint strength. Figure 4 shows the dimensions of hollow shaft specimens. Solid shafts were also manufactured. The adherends were made of steel(S45C,JIS) and aluminum(A5056,JIS). Figure 5 shows the position of strain gauges attached to the outer surface of adherends. The solid and hollow shafts were bonded by an epoxide adhesive(SUMITOMO 3M, Co., Ltd., in Japan, Scotch Weld 1838). Curing was carried out at 60 ℃.

Figure 6 shows the experimental apparatus for four−point bending tests. When a force W is applied to adhesive butt joints by a material testing machine, a bending moment M (= W/2 × 50) is created. The magnitude of the force W is measured with a load cell, the strain in circumferential direction is measured by strain gauges attached to the shaft, and the output signals are recorded by an X−Y recorder through dynamic amplifiers.

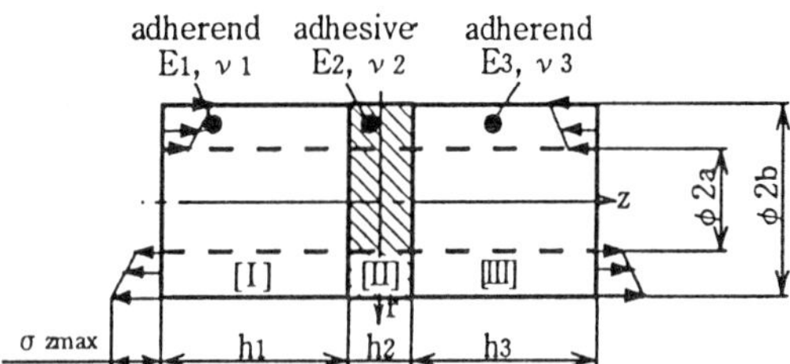

Fig.1 Adhesive butt joint of dissimilar hollow shaft subjected to external bending moment

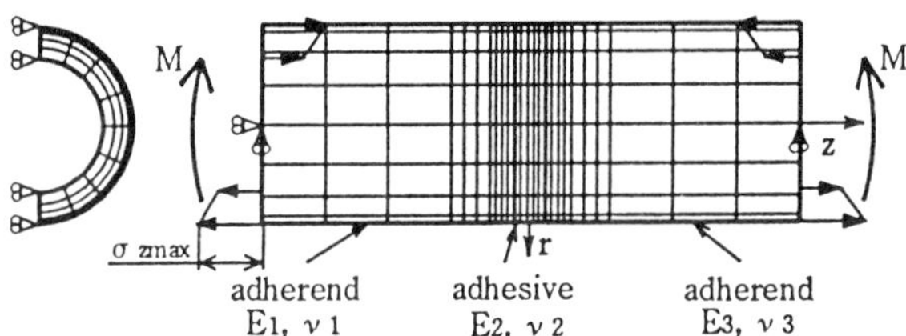

Fig.2 Boundary conditions and an example of mesh divisions in FEM analysis

Table 1 Mechanical property of adherends and adhesive

	adherend (St)	adherend (Al)	adhesive (epoxy)
Young's modulus E[GPa]	205.8	71.5	3.63
Poisson's ratio ν	0.30	0.33	0.37
Yield stress σ_Y[MPa]	425.3	261.7	21.2
c[GPa]	19.8	9.00	0.0304

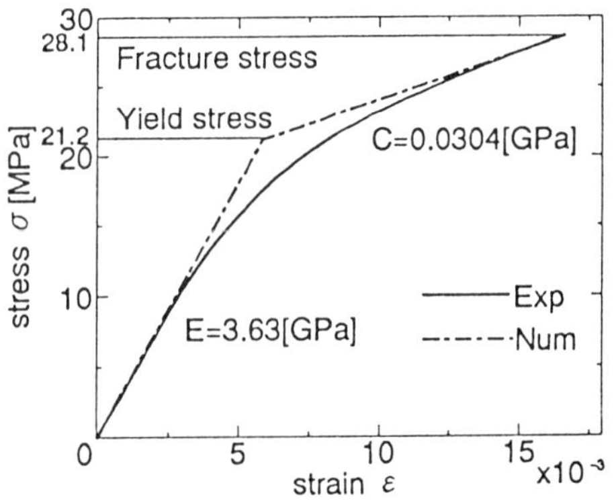

Fig.3 Stress−strain diagram (epoxy adhesive)

4. Results of Analysis and Comparison with Experimental Data

4.1 Results of Analysis

Figure 7 shows the effect of the ratio of Young's moduli for the adherends to that for the adhesive, E_1/E_2, on the stress distributions at the interface($\theta =3/2\cdot \pi$), where the values E_1/E_2 are chosen as 1, 20 and 57 and the adhesive butt joint is made of similar substrate materials ($E_1 =E_3$). The ordinate is the normalized maximum principal stress σ_1 / σ_{zmax}, where σ_{zmax} is shown in the model of Fig. 1 and the abscissa is the ratio of the distance from the center to the inner radius of the finite hollow cylinder. The stress distributions at the interface are shown at the angle $\theta =3/2\cdot \pi$. This figure shows the stress distribution at the interface where the element stress exceeds the yield stress and reaches the fracture stress of the adhesive near the point of r=b and $\theta =3/2\cdot \pi$. Furthermore, the analysis was carried out by releasing the nodes of the yielded elements. The yielded areas which are increased with an increase of the external bending moment are indicated by the black color in this figure.

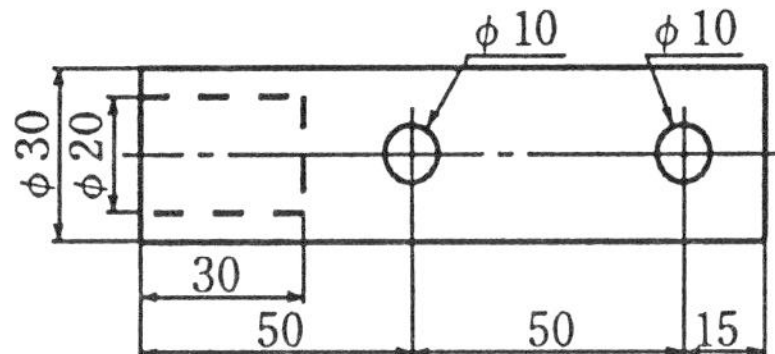

Fig.4 Geometry and the dimentions of solid and hollow shaft specimens used in experiments

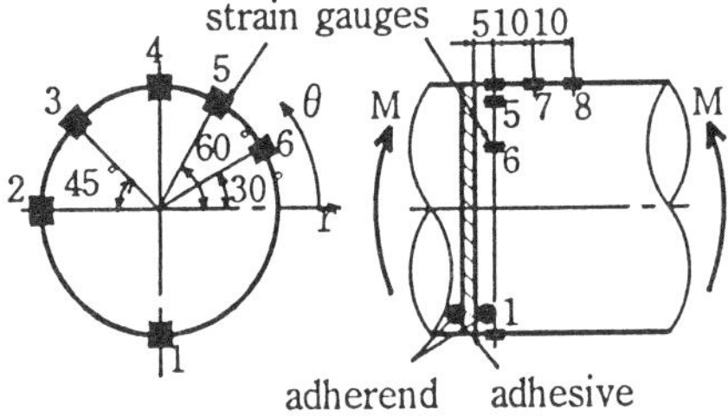

Fig. 5 Attached positions of strain gauges

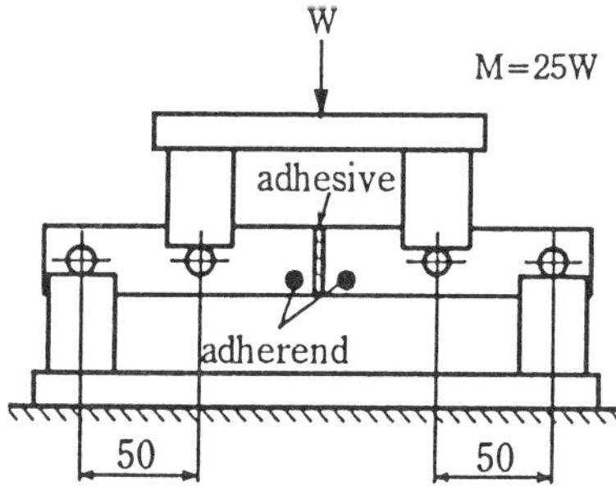

Fig.6 A schematic of experimental apparatus

Figure 7 reveals that the maximum principal stress, σ_1, increases near the outer diameter r/b=1.5 as the value E_1/E_2 approaches 1.

Figure 8 shows the stress distribution at the interface, when the ratio of Young's moduli of adherends is held constant at $E_1/E_3 =2.9$(St−Al) and $E_1/E_2 =57$. The symbol I indicates the interface with the adherend of larger Young's modulus and the symbol III indicates the opposite side interface. It is seen that the normalized maximum principal stress distribution at the position of r/a=1.5 and $\theta =3/2\cdot \pi$ increases with an increase of the value E_1/E_2. In the case of dissimilar shafts, it is predicted that the joint fracture occurs at the interface joining the adherend with smaller Young's modulus because the principal stress is larger than that at the interface joining the adherend with larger Young's modulus. This result is different than the failure behavior of similar joints subjected to a tensile load (Sawa et al., 1987). In addition, it is predicted that the joint strength of dissimilar hollow shafts is smaller than that of similar hollow shafts.

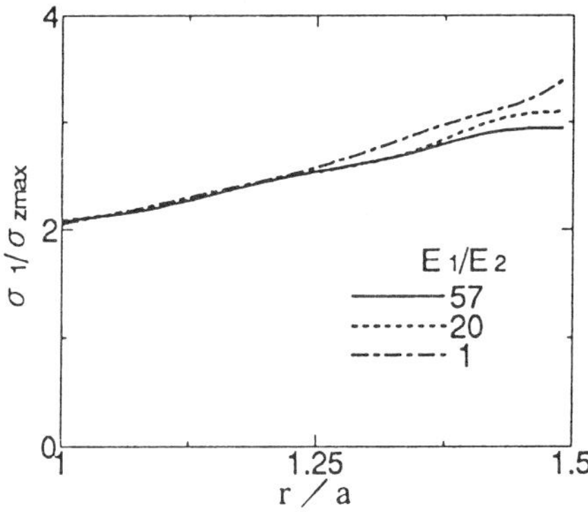

Fig.7 Effects of the ratio of Young's moduli E_1/E_2 on the maximum principal stress distribution at the interface
$\left(\begin{array}{l} h_2/b=0.003, E_1/E_3=1, \theta =3/2\ \pi, \\ h_1=h_3=50[mm], h_2=0.05[mm] \end{array} \right)$

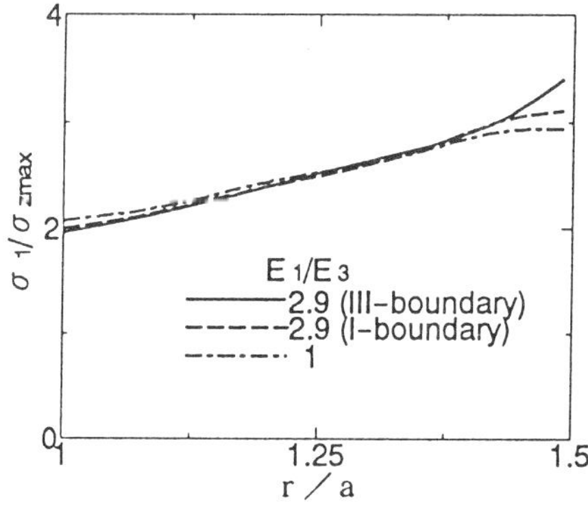

Fig.8 Effects of the ratio of Young's moduli E_1/E_3 on the maximum principal stress distribution at the interface
$\left(\begin{array}{l} h_2/b=0.003, E_1/E_2=57, \theta =3/2\ \pi, \\ h_1=h_3=50[mm], h_2=0.05[mm] \end{array} \right)$

Figure 9 shows the effect of the adhesive thickness on the stress distribution at the interface. The normalized maximum principal stress distribution near the edge of outer diameter(r/a=1.5) increases with a decrease of h_2/b. This result is opposite of that for joints subjected to external tensile loads. In order to investigate the effects of tensile loads on the interface stress distributions, FEM analysis was performed. Figure 10 shows an example of mesh division in FEM for the adhesive joint subjected to tensile loads. The joint geometry is similar to that subjected to bending moments. Figure 11 shows the effect of the ratio of Young's moduli for the adherends to that for the adhesive E_1/E_2 on the maximum principal stress at the interface($\theta = 3/2 \cdot \pi$). Figure 12 shows the effect of the adhesive thickness on the maximum principal stress at the interface. In the case where joints are subjected to tensile loads, a uniform stress distribution of σ_{zmax}, which is the maximum value in the case of bending moments, acts on the ends of joints. From the comparisons between the cases of tensile loads and bending moments, it is seen that the results in the case where external bending moments are applied are opposite to those in the case where tensile loads are applied. We note that when joints are subjected to bending moments, the maximum principal stress σ_{1max} increases as the value E_1/E_2 approaches 1 and the value h_2/b decreases.

Figure 13 shows the changes in interface stress distribution with increasing bending moments. When the principal stress of an element reaches the fracture stress of the adhesive, the analysis is then carried out by releasing the corresponding nodes. The interface stress distributions within the elastic region are also identified. From a comparison of the interface stress distributions for a solid shaft and a hollow shaft for the same outer diameter, the difference in the maximum principal stresses is small but the difference in the slope of stress distributions is very different. Figure 14 shows the variation of the yield region at the interface of the adherend with smaller Young's modulus as the bending moment is increased. In this paper, joint fracture is defined when the stress at the half part of the interface reaches the fracture stress. The difference of bending moments between when yield of adhesive starts and when joint fracture occurs was very small. The region of yielded part progresses toward the center from the edge of interface with an increase of the bending moment.

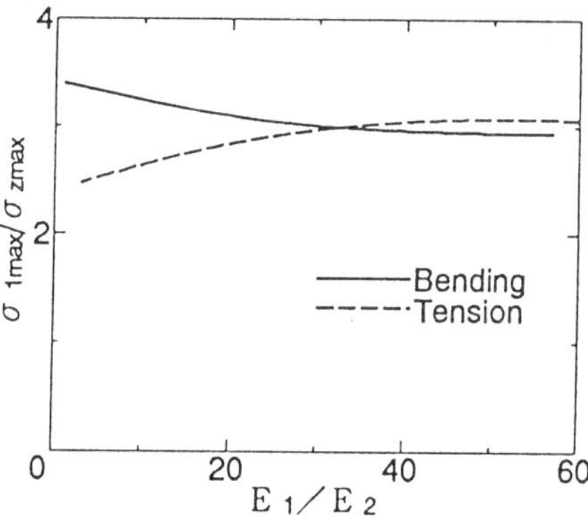

Fig.11 Effects of the ratio of Young's moduli E_1/E_2 on the maximum principal stress in both cases where bending moments and tensile loads are applied
$$\begin{pmatrix} h_2/b=0.003, E_1/E_3=1, \theta = 3/2 \pi, \\ h_1=h_3=50[\text{mm}], h_2=0.05[\text{mm}] \end{pmatrix}$$

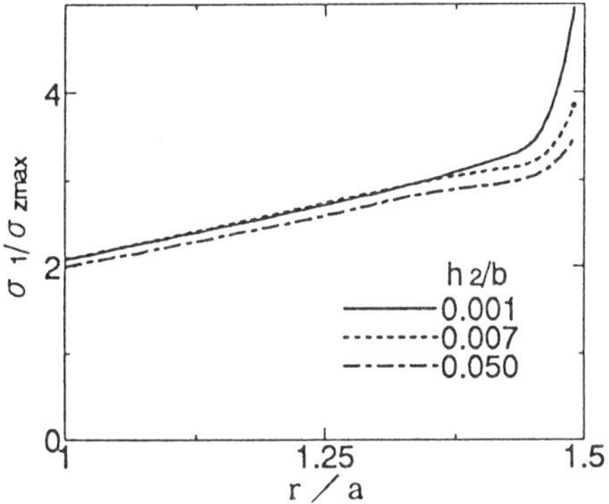

Fig.9 Effects of the adhesive thickness on the maximum principal stress distribution at the interface
($E_1/E_2=57, E_1/E_3=1, \theta = 3/2 \pi, h_1=h_3=50[\text{mm}]$)

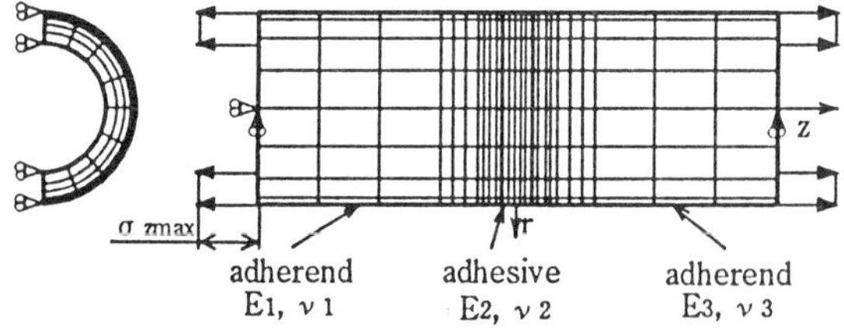

Fig.10 Boundary conditions and an example of mesh divisions in FEM analysis when a tensile load is applied to an adhesive butt joint

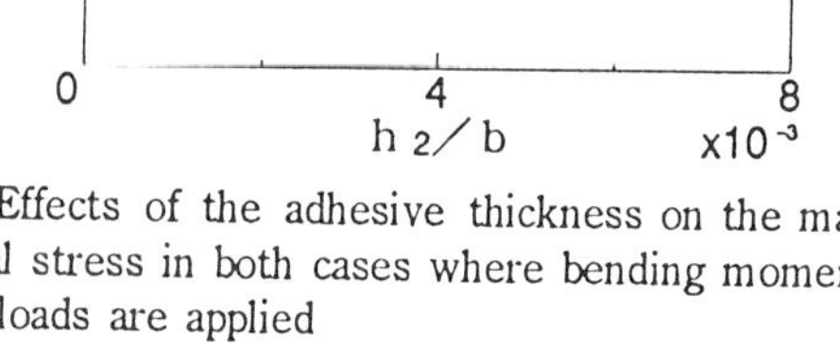

Fig.12 Effects of the adhesive thickness on the maximum principal stress in both cases where bending moments and tensile loads are applied
($E_1/E_2=57, E_1/E_3=1, \theta = 3/2 \pi, h_1=h_3=50[\text{mm}]$)

4.2 Comparisons between Analytical and Experimental Results Concerning Strain

Figure 15 shows the comparison between the measured and the numerical results concerning strain in adherends. The ordinate represents the strain ε_z and the abscissa is the position θ of the attached strain gauge. Taking into consideration the joint symmetry, the range of angle is taken as $0 \leqq \theta \leqq 90$. Figure 15 reveals a fairly good agreement between the numerical and the experimental results.

4.3 Comparisons Between Analytical and Experimental Results on Joint Strength

The bending moment corresponding to joint fracture is identified when half part of the interface reaches the fracture stress of the adhesive. Table 2 shows bending moments obtained by elastoplastic analysis(E−P), elastic analysis(E) and experimental results when fracture is occurred. The experimental results are the average values of 20 measurements in four−point bending tests. The standard deviation of the experimental results is between 1.40 and

1.96[kN·mm]. In the elastic finite element analysis, the joint fracture is defined such that the maximum principal stress at the edge of interface reaches the fracture stress obtained from cylinder specimen tests (Temma et al., 1993) under tensile loading and the bending moment is defined as fracture strength of the joint. The fracture stress of cylinder specimen for St−St, Al−Al and St−Al were measured as 47.4[MPa], 43.9[MPa] and 41.2[MPa], respectively. The adhesive joint strength of St−St is greater than that of Al−Al adhesive joint in the similar shafts. This result agrees with the predicted values shown in Fig. 7. It is found that the strength of similar shafts is greater than that of dissimilar shafts. This result agrees with the predicted values in Fig. 8. From comparisons between the case of hollow shafts and the case of the solid shaft of which the outer diameter is the same as hollow shafts, it is found that the joint strength of solid shafts is greater than that of hollow shafts. The predicted value of the bending moment for joint fracture obtained by elastoplastic finite element analysis(E−P) is in fairly good agreement with the experimental one. Concerning elastic finite element analysis(E), the value of bending moment is smaller than that

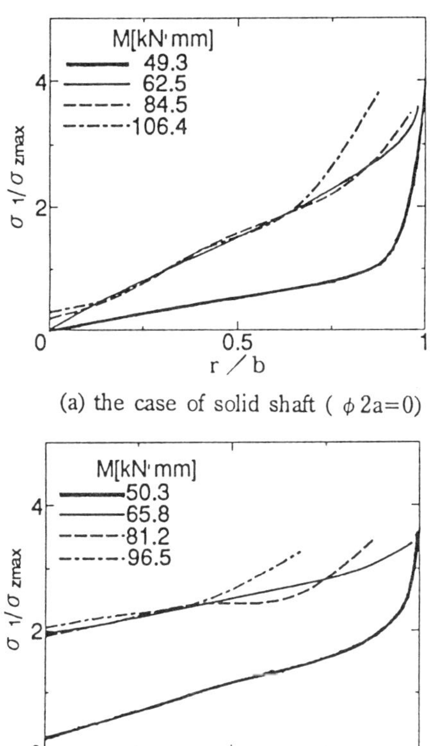

(a) the case of solid shaft ($\phi 2a=0$)

(b) the case of hollow shaft (b/a=1.5)

Fig.13 Changes in distribution at the interface with an increase of bending moment
$\left(\begin{array}{l} h_2/b=0.003, E_1/E_2=57, E_1/E_3=2.9,\ \theta =3/2\ \pi, \\ h_1=h_2=50[mm], h_2=0.05[mm] \end{array} \right)$

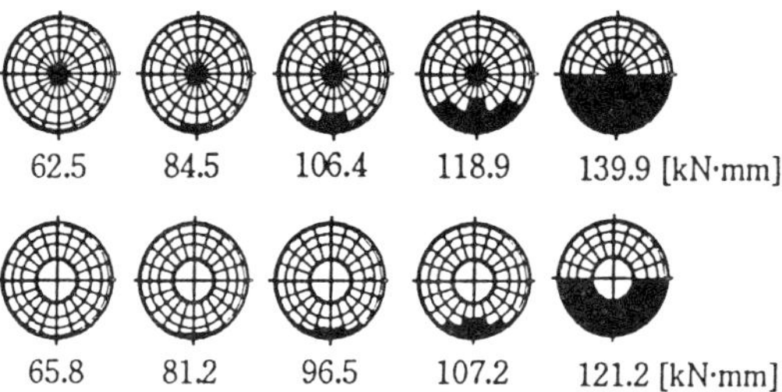

Fig.14 Changes in yield region at the interface with an increase of bending moment
$\left(\begin{array}{l} h_2/b=0.003, E_1/E_2=57, E_1/E_3=2.9 \\ h_1=h_3=50[mm], h_2=0.05[mm] \end{array} \right)$

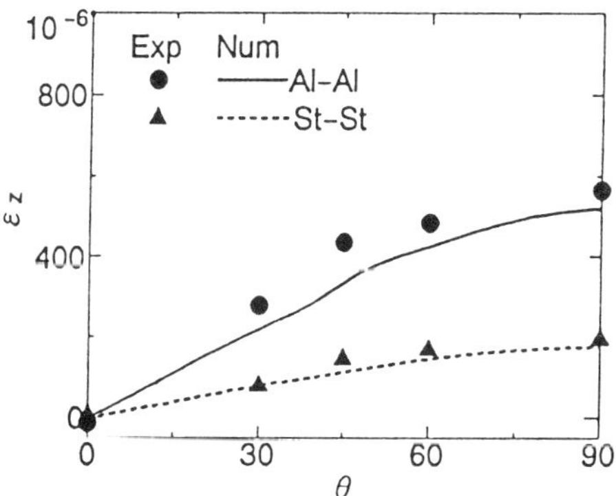

Fig.15 Comparisons between the numerical and the experimental results concerning strain of adherend
$\left(\begin{array}{l} \phi 2a=20[mm],\ \phi 2b=30[mm], h_1=h_3=50[mm], \\ h_2=0.05[mm], M=98[kN·mm] \end{array} \right)$

of elastoplastic analysis. It was found experimentally that the joint fracture occurred at the interface. In the case of dissimilar shafts, the joint fracture occurred at the interface of the aluminum adherend. This result agrees with elastoplastic analysis.

5. Conclusions

This paper dealt with the interface stress distributions and the strength evaluation of adhesive butt joints of dissimilar hollow shafts subjected to external bending moments. The following results are obtained.

(1) The interface stress distributions of adhesive butt joints consisting of dissimilar hollow shafts and subjected to external bending moments were analyzed by elastoplastic finite element analysis, and the progression of the yield area at the interface was indicated. As a result, it is found that the maximum principal stress at the interface of the adherend with smaller Young's modulus is larger than that at the interface of the adherend with larger Young's modulus. In addition, it is seen that the maximum principal stress increases with a decrease of the ratio of Young's moduli for the adherend to that for the adhesive, E_1/E_2, and with a decrease of the adhesive thickness.

(2) Adhesive butt joints of dissimilar hollow shafts subjected to tensile loads were also analyzed. As a result, it is found that the analyzed characteristics of joints subjected to tensile loads is opposite of joints subjected to bending moments.

(3) Using the interface stress distributions, joint strength is estimated. As a result, it is found that the joint strength of dissimilar hollow shafts is smaller than that of similar hollow shafts. The joint strength of solid shafts of which the outer diameter is the same as hollow shafts is estimated. It is found that the joint strength of solid shafts is larger than that of hollow shafts. In addition, it is found that the joint strength of similar shafts increases with an increase of the ratio of Young's moduli for the adherend to that for the adhesive E_1/E_2.

(4) Concerning strain of adherends, the numerical results are in fairy good agreement with the experimental results. In addition, fairly good agreement is observed between the numerical and the experimental results concerning joint strength. It is found experimentally that the joint strength of dissimilar hollow shafts is smaller than that of similar hollow shafts.

References

Wah. T., 1973, "Stress Distribution in a Bonded Anisotropic Lap Joint", Trans. ASME, Journal of Engineering Materials and Technology, pp. 174−181.

Table 2 Comparisons between the numerical and the experimental results concerning joint strength (bending moment : M)

(a) Case of hollow shaft [kN·mm]

	St−St	Al−Al	St−Al
Exp	156	143	125
Num(E−P)	154.2	139.9	121.2
Num(E)	139.1	110.9	106.0

(b) Case of solid shaft [kN·mm]

	St−St	Al−Al	St−Al
Exp	175	162	143
Num(E−P)	174.3	158.9	139.9
Num(E)	171.6	140.2	130.8

R. D. Adams and N. A. Peppiatt, 1974, "Stress Analysis of Adhesive−bonded Lap Joints", Journal of Strain Analysis, 9−3, pp. 185−196.

Ming−Yi Tsai and J. Morton, 1994, "A Note on Peel Stresses in Single−Lap Adhesive Joints", Trans. ASME, Ser. E, 61, pp. 712−715.

T. Sawa, et al, 1987, "Stress Analysis of T−type Butt Adhesive Joint Subjected to External Bending Moment", Journal of the Adhesion, 24−1, pp. 1−15.

K.Temma, et al, 1990, "Two−Dimensional Stress Analysis of Adhesive Butt Joints Subjected to Cleavage Loads", Int. J. Adhesion and Adhesive, 10−4, pp. 285−293.

T. Sawa, et al, 1990, "Three−Dimensional Stress Analysis of Adhesive Butt Joints of Solid Cylinders Subjected to External Tensile Loads", Journal of Adhesion, 31, pp. 33−34

T. Sawa, et al, 1989, "Axisymmetric Stress Analysis of Adhesive Butt Joints of Dissimilar Solid Cylinders Subjected to External Tensile Loads", Int. J. Adhesion and Adhesive, 9−3, pp. 161−169.

K. Temma. et al, 1994, "Two−Dimensional Stress Analysis and Strength of Band Adhesive Butt Joints of Dissimilar Adherends Subjected to External Bending Moments", JSME International Journal, 37−3, pp.246−254.

Y. Nakano, et al, 1991, "A Stress Analysis of Adhesive Butt Joints of Dissimilar Materials Subjected to Cleavage Loads", Journal of Adhesion, 34, pp. 137−151.

T. Sawa et al, 1987, "Three−Dimensional Stress Analysis of Adhesive Butt Joints Subjected to Tensile Loads(The Case where Adherends are Two Hollow Cilynders)", Trans. JSME (in Japanese), 53−492, pp. 1685−1691.

K.Temma, et al, 1993, "A Two−Dimensional Stress Analysis and Strength Evaluation of Adhesive Butt Joints of Dissimilar Adherends to Tensile Loads", Adhesion Soc. Japan (in Japanese), 29−7, pp. 302−315

**DE-Vol. 87, Reliability, Stress Analysis, and Failure Prevention
Issues in Emerging Technologies and Materials
ASME 1995**

A COMPARISON OF ADHESIVELY BONDED SINGLE LAP, SCARF AND BUTT JOINTS

Jeffrey S. Baylor and Erol Sancaktar
Clarkson University
Center for Advanced Materials Processing
Department of Mechanical and Aeronautical Engineering
Potsdam, New York 13676-5729

ABSTRACT

The commercially available ANSYS finite element code is used to execute a two-dimensional stress analysis of single lap, scarf, and butt joint geometries. The effect of the scarf angle on the normal, shear and transverse stress peaks and distributions is assessed for design purposes. The single lap and butt joint geometries are considered as special cases of scarf joints with scarf angles of 0° and 90°, respectively. Cases with 25.4 mm adhesive layer length are considered along with 30°, 45°, 60°, and 75° scarf angled joints of 25.4 mm thick substrates. In the latter case the adhesive layer length becomes a function of the scarf angle since the substrate thickness is constant. Hence, the effect of the adhesive layer length on the stress distributions for this class of joints is also assessed. Joint efficiencies are evaluated by comparing the ratios of the maximum stresses divided by the average stresses for each mode of stress created by externally applied tensile loads on the substrates.

INTRODUCTION

Applications for adhesive joints can be found in industries ranging from the construction industry to the aerospace industry. As their need to produce more reliable and cost efficient joints increases, there will be an increasing demand for a general design methodology. The methodology is needed to compare the array of adhesive joint configurations and determine which type of joint best suits the design's need and then optimize its geometry. This paper presents an introduction to this process by comparing the stresses, modes of loading and geometries in order to eventually develop a general methodology to compare adhesive joints for design optimization. The specific joints investigated are the single lap joint, scarf joint and the butt joint.

PROCEDURE

The finite element method was used to analyze the stress distributions in the different configurations of adhesive joints. All of the preprocessing, solving and postprocessing was done using the software package: ANSYS version 5.0. The joints were considered to be under a constant load and the state of stress was assumed to be plain stress. This assumption allowed the use of a static, two dimensional model. A "Plane 42" element (2-D 4-node quadrilateral structural solid) was used to construct the model. The quadrilateral elements were used instead of the triangular elements because they yield more accurate results for structural problems.

The material properties of a commercially available adhesive were utilized and the substrates were modeled as aluminum. All of the adhesive material data was obtained from Renieri et. al (1976), and the aluminum material data was obtained from Shigley and Mischke (1989). An analysis of this type requires the modulus of elasticity and the Poisson's ratio. These properties are summarized in Table 1.

There were two general models constructed to analyze the three joint types, since a butt joint is a specific case of the scarf joint where the angle of the adhesive is 90° from the surface of the substrate. All of the geometric quantities were defined by variables so that the model dimensions could be altered with ease. Figures 1 and 2 show the drawings of the scarf and single lap joints with each geometric quantity dimensioned as its variable name. Note that these figures are not to scale and are only meant to clarify the meaning of the variables in the models.

The constraints and loads were applied so as to mimic a tensile load on the joint while it was secured in nonrotating clamps. The load was applied along the axis parallel to the substrate length as a distributed load acting away from the right most surface (the positive X direction). All of the nodes on this surface were coupled in every degree of freedom (X and Y translation and rotation around Z), coupling implies that all of the nodes will move or rotate the same amount in the given degree of freedom. These nodes were constrained from any movement perpendicular to the load. This constraint along with the coupling also constrained these nodes from rotation. The left most surface (the negative X direction) was constrained in every degree of freedom, succeeding in simulating a fixed support. Since there were no curved surfaces, there was no problem in creating a mesh which had the identical geometry of the represented joint.

Finite element results are obtained by interpolating between the nodes of a mesh. The interpolation errors have been investigated by determining the effect of the mesh density on the stress at a point of interest. The point that was chosen was one millimeter (from left to right) from the overlap of the single lap joint along the loaded substrate interface. The mesh was refined across the adhesive and along the length of the interface, the results are shown in Figures 3 and 4. Since the density can vary in two dimensions a "brute force" optimization technique was adopted. Multiple iterations of single lap joint models were executed with different numbers of elements across the adhesive and a different number of elements per millimeter along the adhesive length. The density along the length was chosen to be four elements per millimeter, and five elements were chosen to span the adhesive thickness. The Von Mises stress (SEQV) and the principal stress difference, i.e. twice the maximum shearing stress (SINT) has been plotted for the iterations that were used to make these decisions. Figures 3 and 4 reveal that the change in stress due to the change in mesh density approaches zero, and the densities chosen has been shown to yield results constant to five significant digits.

Figure 3 shows the effect of mesh density variation along the overlap length when there are five elements across the 0.2 mm thick adhesive layer. Both the Von Mises stress and the principal stress difference (i.e. 2 τ_{max}) values are plotted. Figure 3 reveals that repeatable analytical predictions are obtained when a mesh density of four elements per millimeter or higher is used. Consequently, it was decided that four elements per millimeter would be used along the length direction.

Figure 4 shows the effect of mesh density variation across the adhesive thickness when there are four elements per millimeter along the length direction. Both the Von Mises stress and the principal stress difference (i.e. 2 τ_{mzx}) values are plotted. Figure 4 reveals that repeatable analytical predictions are obtained when five elements or more are used across the 0.2 mm thick adhesive layer. Consequently, it was decided that five elements would be used across the adhesive thickness. A similar approach was taken to determine how long the scarf joint substrates had to be in order to sufficiently negate the effects of the constraints and the loading on the adhesive area.

For this purpose distributions of the normal stress (perpendicular to the adhesive layer), transverse stress (along the adhesive layer), and the shear stress were obtained for the adhesive/substrate interfaces on the loaded and fixed (boundary) sides of a 30° scarf joint. Scarf angle of 30° was chosen since it was the case that would be affected most by the substrate length. Results were obtained using the substrate length of 40 mm and also 80 mm to assess the effects of substrate length on stress distributions. The variation in stress distributions with this change of substrate length are shown in Figures 5 through 7 respectively for normal,

transverse and shear stress distributions on the first 4 mm of interface length. As can be seen, the effect of substrate length above 40 mm is negligible in this case.

In order to simplify the problem some assumptions were made. The strains in both the substrates and the adhesive were considered linear and the stresses were assumed to be acting in one plane. The bond between the substrates and the adhesive was assumed to be perfect, and the adhesive and the interface was assumed to be free of voids. These assumptions, however, will not invalidate the comparative use of these models.

Two specific scarf joint variations were looked at. First, the adhesive was maintained at 25.4 mm in length, and second, the substrate thickness was maintained at 25.4 mm. In each of these situations, the angle of the adhesive joint (theta) was set to 30°, 45°, 60°, 75°, and 90°. The single lap joint model was analyzed using the identical parameters, but with a substrate thickness of 12.7 mm. These results will be compared with the 30°, 60°, and 90° scarf joints. The geometric quantities that remained constant throughout the iterations and those that were dependant on the scarf angle theta have been tabulated in Table 2. The applied load was equivalent to a one Newton point force.

RESULTS

All of the stress paths to be presented here coincide with the interfaces of the adhesive and the substrates. The coordinate system in which the results will be presented has been rotated by the angle theta. This means that the results X-axis will lie along the substrate interface and the results Y-axis will be normal to the interface.

The results of finite element analyses on single lap and scarf joints revealed that the distributions of normal, σ_y, transverse, σ_x, and shear τ_{xy} stresses on the loaded and fixed (boundary condition) sides of the adhesive/substrate interfaces are anti-symmetric. In other words, the stress distributions are rotated 180° around a "y" axis going through the adhesive layer at half length. Consequently, we will present results only for the loaded side of the substrate/adhesive interface.

The variations of normal, transverse, and shear stresses are compared between lap, butt, and scarf joints of 30 and 60 degrees taper in Figures 8, 9, and 10 respectively for an adhesive layer length of 25.4 mm. It can be observed that, in general, the distributions of stresses become more uniform along the adhesive layer (x-axis) as one goes from lap joint to butt joint in an increasing order of scarf angle. Note that in this analogy the lap joint is assigned a scarf angle of 0° while the butt joint is assigned 90° scarf angle.

Similar behavior is observed when normal, transverse, and shear stress distributions are compared for scarf angle increments of 15° between 30° and 90° (butt joint) (Figures 11, 12, 13, respectively). Note that for the distribution of the shear stress, the butt joint appears to have more nonuniformly than those for 60° or 75° scarf angles. The magnitudes of the shear stress for the butt joint, however, are much smaller than those for the 60° or 75° scarf angles.

The maximum stress values observed in Figures 8 through 13 are tabulated in Table 3 along with average stress values and the ratios of maximum divided by the average for each of transverse, σ_x, normal, σ_y, and shear, τ_{xy}, stresses. It can be observed that the ratios for all three stress modes (i.e., transverse, normal, and shear) are reduced as the scarf angle is increased from 0° (lap joint) to 60°. Any relative increases in the stress ratios when the scarf angle is increased beyond 60° and up to 90° (butt joint) can be considered negligible.

The variation of peak stresses tabulated in Table 3 with respect to the scarf angle is shown in Figure 14. Note that the magnitude of the peak normal stress is minimum when we use 30° scarf joint. The magnitudes of transverse and shear stresses, on the other hand, are minimum when we use the butt joint.

When design situations require a constant substrate thickness (see Figure 2), such as 25.4 mm, for the scarf joints then the use of smaller scarf angles would require longer adhesive layer lengths until we reach single lap configuration at 0° scarf angle. In these situations the length of the adhesive layer would be (see Figure 2):

$$ALE = STH/\sin\theta.$$

Table 4 shows the maximum, average and maximum to average ratios for the transverse, normal, and shear stresses in 30°, 45°, 60° and 75° scarf joints. Comparison with Table 3 reveals that all stress magnitudes are reduced with increasing adhesive layer length. Furthermore, all stress ratios are also reduced except for the transverse stress ratio for the 75° scarf, which remains the same, and the normal stress ratio for the 30° scarf which shows a 25% increase in $\sigma_{XMAX}/\sigma_{TAVG}$ with 100% increase in adhesive length. Note, however, that for this last case of 30° scarf, the σ_{XMAX} is reduced by 64% and the σ_{YAVG} is reduced by 71%.

Based on Goland and Reissner's work (1944) it is also known that, for the single lap geometry, increases in the adhesive layer length results in more uniform stress distributions and reductions in stress peaks.

The trends and magnitudes that appear in the data can be explained using strength of materials and applied elasticity. For example, using strengths of materials, the average normal stress for the butt joint should equal the load divided by the cross sectional area. The load is equivalent to 1 N and the area is equal to the substrate thickness (25.4 mm) times a unit thickness (1 mm). The result is 39.37 kPa which compares well with the average normal stress of 39.41 kPa from the model (see Table 3).

For the single lap joint the shearing stress should be equal to the load divided by the area and acting in the negative X direction on the positive Y surface, making it a negative shearing stress. The load is 1 N and the area is 25.4 mm times a unit thickness (1 mm). The theoretical result is - 39.37 kPa which compares reasonably to what the model yielded as the average shearing stress, -38.4 kPa.

CONCLUSIONS

The finite element analysis revealed that, in general, the distributions of stresses become more uniform along the adhesive layer as one goes from lap joint to butt joint in an increasing order of scarf angle when the substrate is loaded in tension. In general, the joint efficiency increases as the scarf angle is increased as revealed by reductions in the ratios of the maximum stresses to average stresses. Increases in the adhesive layer length result in reductions in stress magnitudes as well as reductions in most stress ratios.

REFERENCES

Renieri, M.P., Herakovich, C.T., and Brinson, H.F., 1976, "Rate and Time Dependent Behavior of Structural Adhesives," VPI-E-76-7, Blacksburg, Virginia, p. 39.

Shigley, J.E, and Mischke, C.R., 1989, "Mechanical Engineering Design," 5th ed., McGraw-Hill, New York, p. 729.

Goland, M., and Reissner, E., 1944, J. Applied Mech., Vol. 77, pp. 17-27.

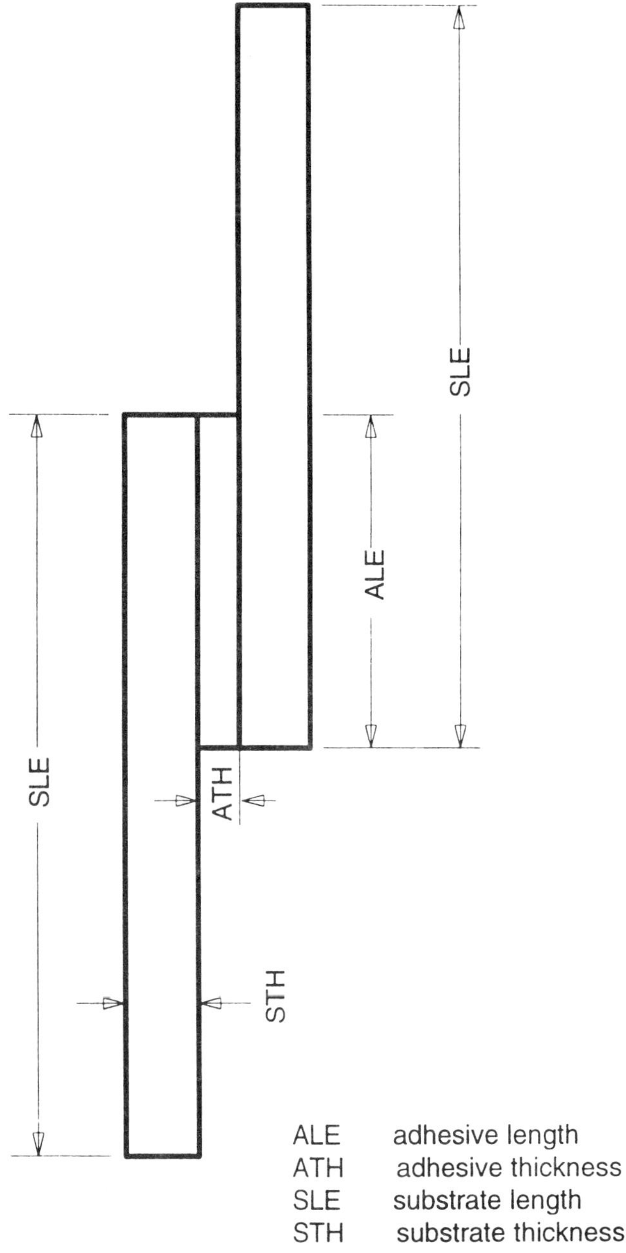

FIGURE 1. THE SINGLE LAP GEOMETRY

TABLE 1. MATERIAL PROPERTIES

	Young's Modulus	Poisson's Ratio
Aluminum	71.00 GPa	0.334
Adhesive	2.015 GPa	0.368

TABLE 2. GEOMETRIC QUANTITIES

	Single Lap Joint	Scarf joint with const. ALE	Scarf joint with const. STH
Substrate length (SLE)	101.6 mm	40.0 mm	40.0 mm
Substrate thickness (STH)	12.7 mm	12.7 to 25.4 mm (theta dependent)	25.4 mm
Adhesive length (ALE)	25.4 mm	25.4 mm	50.8 to 25.4 mm (theta dependent)
Adhesive thickness (ATH)	0.20 mm	0.20 mm	0.20 mm

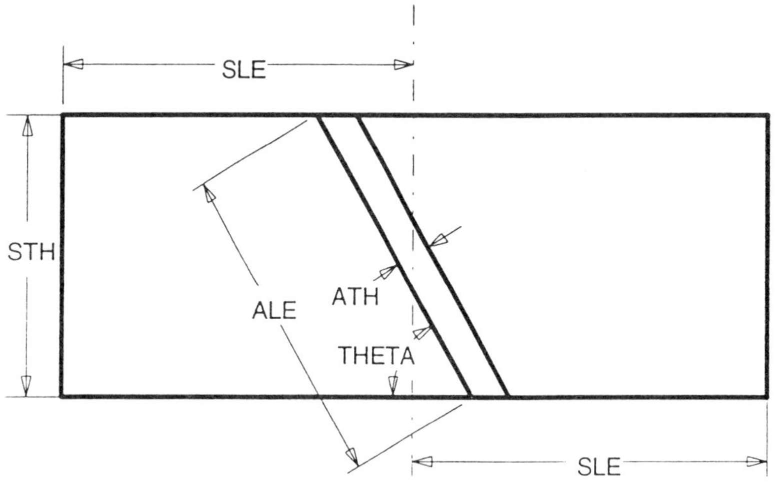

ALE   adhesive length
ATH   adhesive thickness
SLE   substrate length
STH   substrate thickness
THETA angle from substrate surface

FIGURE 2. THE SCARF JOINT GEOMETRY

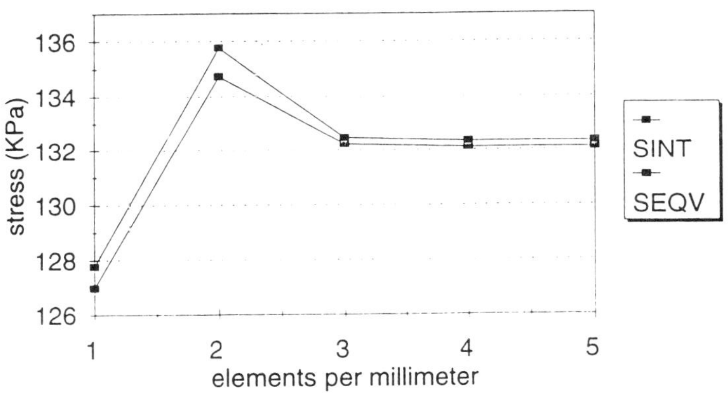

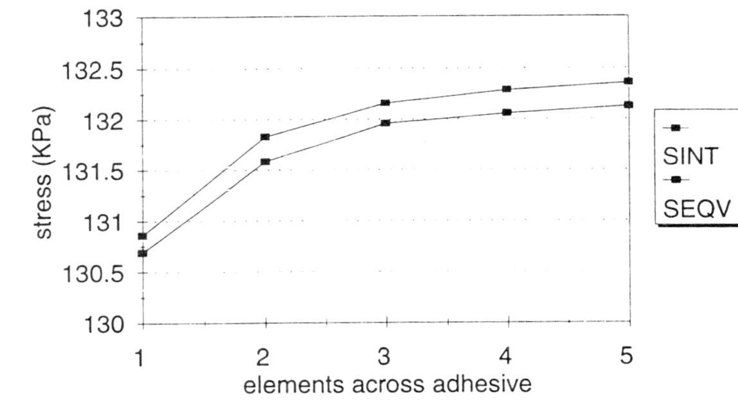

FIGURE 3. THE EFFECT OF MESH DENSITY VARIATION ALONG THE OVERLAP

FIGURE 4. THE EFFECT OF MESH DENSITY VARIATION ACROSS THE ADHESIVE THICKNESS

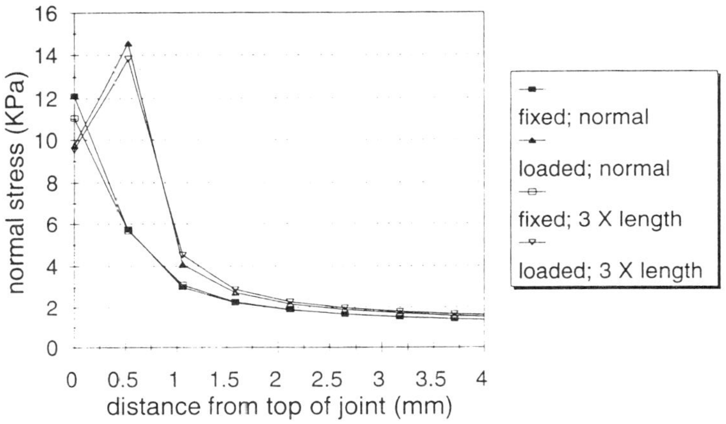

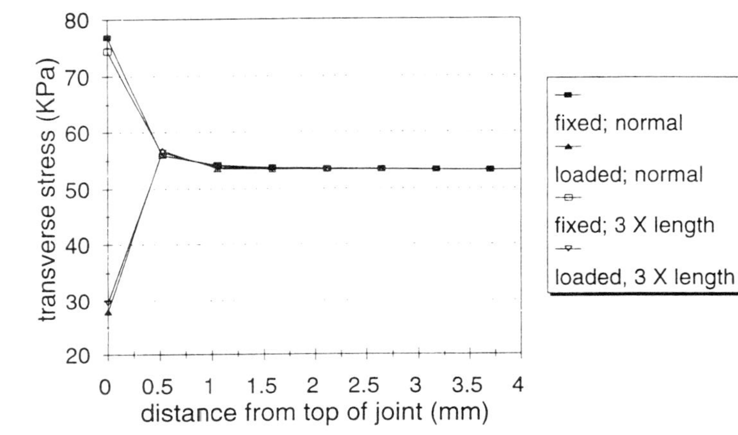

FIGURE 5. THE EFFECT OF SUBSTRATE LENGTH ON THE INTERFACIAL NORMAL STRESS

FIGURE 6. THE EFFECT OF SUBSTRATE LENGTH ON THE INTERFACIAL TRANSVERSE STRESS

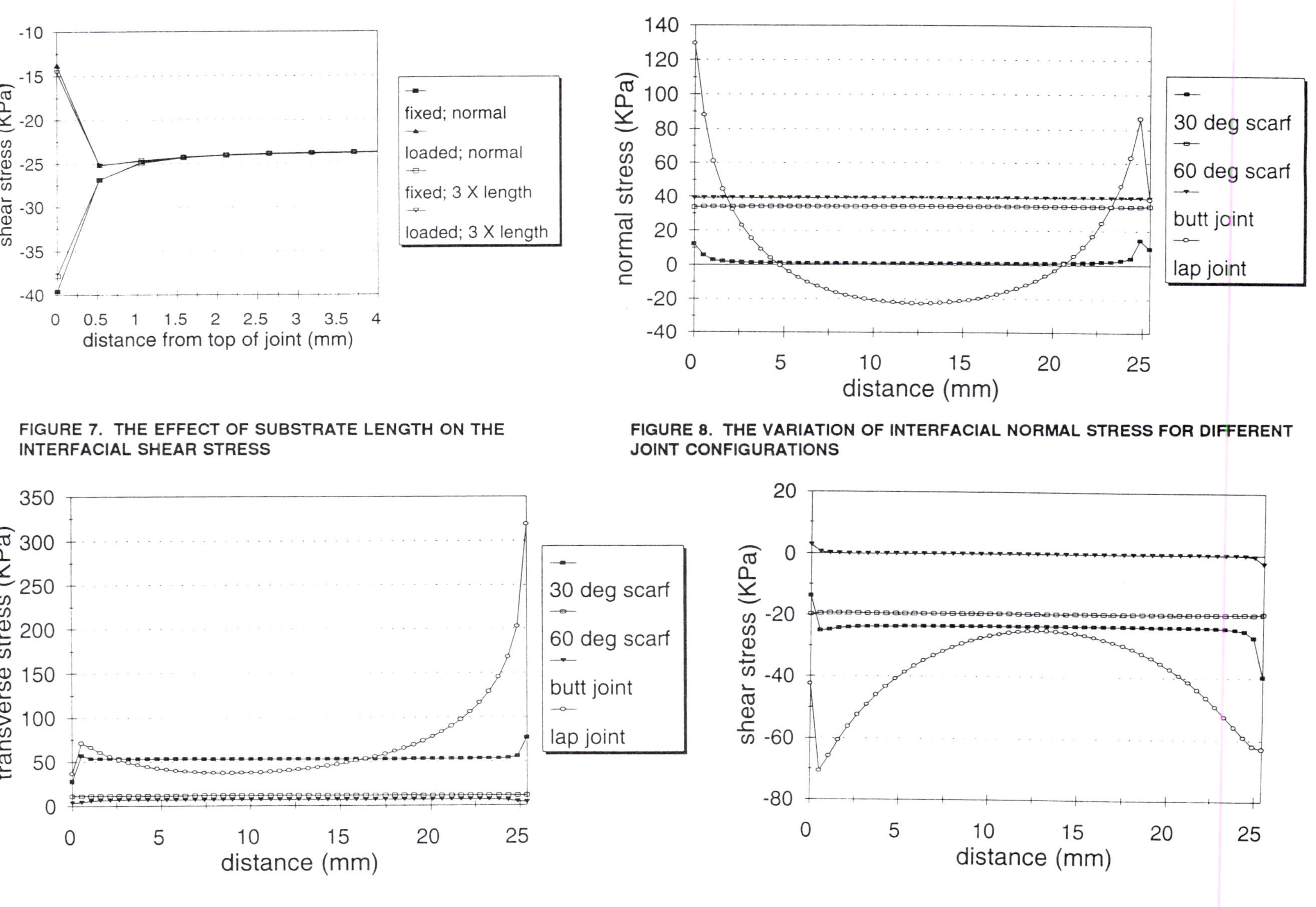

FIGURE 7. THE EFFECT OF SUBSTRATE LENGTH ON THE INTERFACIAL SHEAR STRESS

FIGURE 8. THE VARIATION OF INTERFACIAL NORMAL STRESS FOR DIFFERENT JOINT CONFIGURATIONS

FIGURE 9. THE VARIATION OF INTERFACIAL TRANSVERSE STRESS FOR DIFFERENT JOINT CONFIGURATIONS

FIGURE 10. THE VARIATION OF INTERFACIAL SHEAR STRESS FOR DIFFERENT JOINT CONFIGURATION

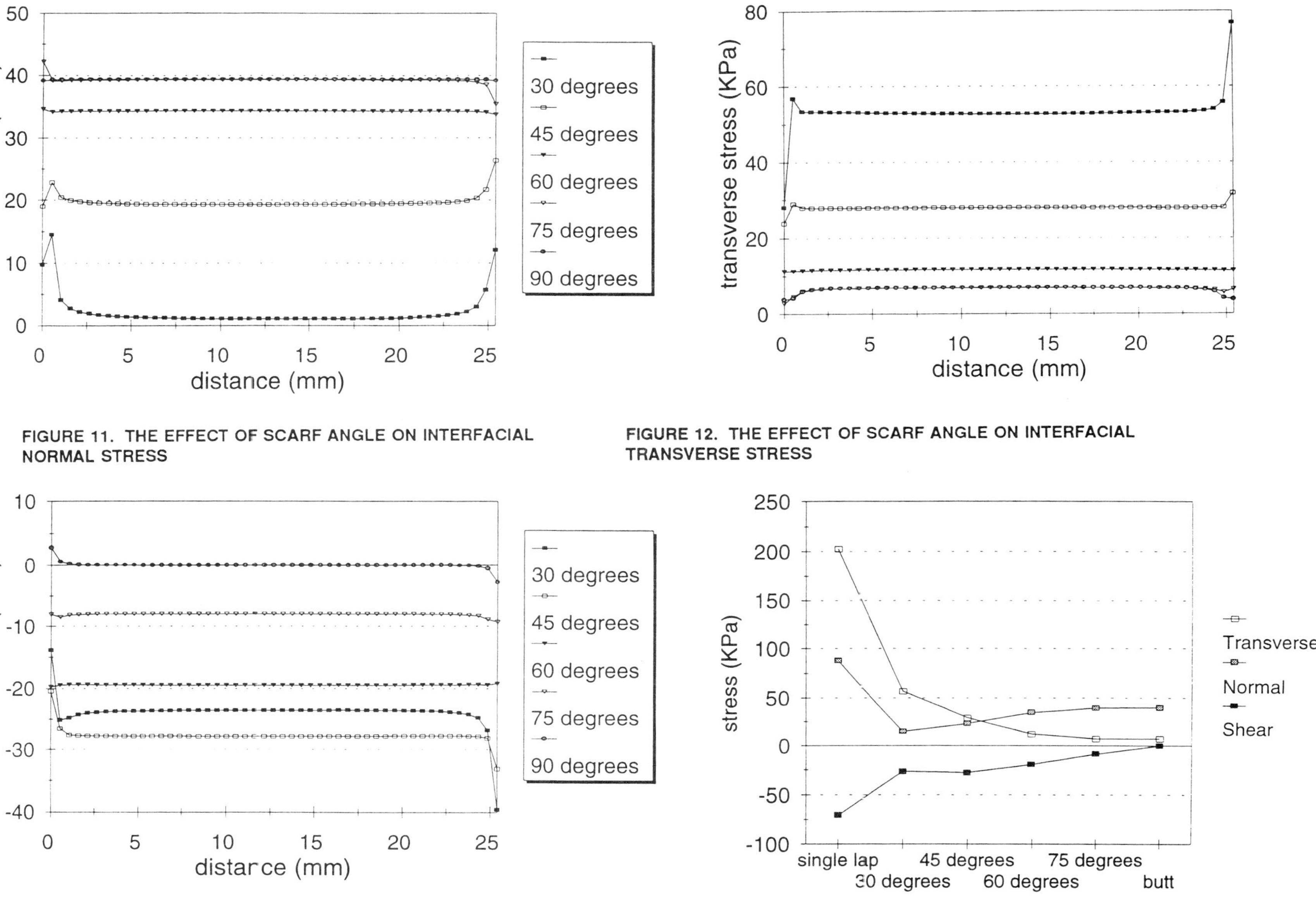

FIGURE 11. THE EFFECT OF SCARF ANGLE ON INTERFACIAL NORMAL STRESS

FIGURE 12. THE EFFECT OF SCARF ANGLE ON INTERFACIAL TRANSVERSE STRESS

FIGURE 13. THE EFFECT OF SCARF ANGLE ON INTERFACIAL SHEAR STRESS

FIGURE 14. VARIATION OF PEAK STRESSES WITH THE SCARF ANGLE

TABLE 3. THE EFFECT OF SCARF ANGLE ON INTERFACIAL STRESSES FOR 25.4 MM WIDE ADHESIVE LAYER

JOINT TYPE	TRANSVERSE STRESS σ_X(kPa)			NORMAL STRESS σ_Y (kPa)			SHEAR STRESS $|\tau_{XY}|$ (kPa)		
	Maximum σ_{XMAX}	Average σ_{XAVG}	Ratio $\sigma_{XMAX}/\sigma_{XAVG}$	Maximum σ_{YMAX}	Average σ_{YAVG}	Ratio $\sigma_{YMAX}/\sigma_{YAVG}$	Maximum $|\tau_{XYMAX}|$	Average $|\tau_{XYAVG}|$	Ratio $\tau_{XYMAX}/\tau_{XYAVG}$
Single Lap	318.62	66.03	4.83	129.72	3.58	36.23	70.68	38.38	1.84
30° Scarf	76.78	53.21	1.44	14.56	1.98	7.35	39.59	23.86	1.66
45° Scarf	31.74	27.90	1.14	26.39	19.68	1.34	33.13	27.76	1.19
60° Scarf	11.74	11.64	1.01	34.65	34.35	1.01	19.72	19.46	1.01
75° Scarf	6.90	6.67	1.03	42.21	39.27	1.07	9.25	8.04	1.15
Butt Joint	6.94	6.62	1.05	39.43	39.41	1.00	2.80	0	N.A.

TABLE 4. THE EFFECT OF SCARF ANGLE ON INTERFACIAL STRESSES FOR 25.4 MM THICK SUBSTRATES

JOINT TYPE	ADHESIVE LENGTH (mm)	TRANSVERSE STRESS σ_X (kPa)			NORMAL STRESS σ_Y(kPa)			SHEAR STRESS $|\tau_{XY}|$(kPa)		
		Maximum σ_{XMAX}	Average σ_{XAVG}	Ratio $\sigma_{XMAX}/\sigma_{XAVG}$	Maximum σ_{YMAX}	Average σ_{YAVG}	Ratio $\sigma_{YMAX}/\sigma_{YAVG}$	Maximum $|\tau_{XYMAX}|$	Average $|\tau_{XYAVG}|$	Ratio $\tau_{XYMAX}/\tau_{XYAVG}$
30° Scarf	50.80	36.93	26.42	1.40	5.24	0.57	9.19	18.81	11.65	1.61
45° Scarf	35.92	22.27	19.75	1.13	14.97	13.73	1.09	23.24	19.59	1.19
60° Scarf	29.33	10.21	10.11	1.01	29.94	29.71	1.01	17.05	16.85	1.01
75° Scarf	26.29	6.66	6.45	1.03	40.76	37.93	1.07	8.93	7.77	1.15

**DE-Vol. 87, Reliability, Stress Analysis, and Failure Prevention
Issues in Emerging Technologies and Materials
ASME 1995**

THE EFFECT OF FIBER TYPE ON THE LEVEL OF STRESS CONCENTRATION CREATED IN FILLETED COMPOSITE RECTANGULAR BARS IN BENDING

Mustafa Gür and Aydın Turgut
Department of Mechanical Engineering
Fırat University
Elazığ, Turkey

Erol Sancaktar
Clarkson University
Center for Advanced Materials Processing
Department of Mechanical and Aeronautical Engineering
Potsdam, New York 13676-5729

ABSTRACT

The effect of fiber type on the level of stress concentration created by fillets were investigated for unidirectional rectangular laminates loaded on or off-axis in bending. The finite element method was used for this purpose utilizing 4-node quadrilateral elements under plane stress conditions. Analyses were performed for three different fiber reinforcements of epoxy matrix: glass, graphite and boron. Typical material properties reported in the literature for glass/epoxy, boron/epoxy and graphite/epoxy were used in calculations.

INTRODUCTION

Fiber-matrix composites especially those with polymeric matrix materials are more susceptible to the deteriorating effects of geometric stress raisers in comparison to isotropic materials. Fiber-matrix interfaces may become the weak link in a composite material when the adhesion forces holding it together deteriorates and fails due to mechanisms such as the absorption of harmful liquids and increases in the stress levels which they transmit (Sancaktar 1990). For example failure of thermoset interfaces due to hydration by water uptake is a known phenomenon. A wicking mechanism has been suggested for liquid transport at such interphases (Sancaktar 1990). Swelling of the polymer matrix due to liquid uptake usually creates stresses in the composite material which may have to be added to the existing stresses. And yet, these stresses may already be high due to the stress concentrations created by geometric discontinuities.

Exposure of fiber ends at a geometric discontinuity or perforation site in a composite structure subjects it to initiation and propagation of cracks which may seriously weaken or fail the entire structure (Agarwal, Broutman 1990).

Due to the reasons cited above designers usually minimize geometric discontinuities and perforations of composite structures. The necessity of using mechanical penetration methods to connect composite structures has been reduced with the advance of solid film epoxies and co-cure techniques. The presence of geometric discontinuities, however, may be unavoidable in many composite structures and, consequently, it would be highly desirable to minimize the stress concentration effects of such discontinuities by optimizing their shapes as well as by choosing and adjusting the composite material properties accordingly.

As the longitudinal as well as transverse elastic moduli values, E_1 and E_2 respectively, for long fiber anisotropic composites depend on the type of fiber used (see Table 1) the stress concentration values due to normal stress:

$$\mathbf{K} = \sigma_{max} / \sigma_o \qquad (1)$$

are expected to depend on the type of fiber used as well as the direction of loading (i.e. on-axis versus off-axis). In equation (1) the maximum stress, σ_{max}, is given by:

$$\sigma_{max} = [(\sigma_x + \sigma_y)/2] + [(\sigma_x + \sigma_y)^2/4 + \tau_{xy}^2]^{1/2} \qquad (2)$$

and the nominal flexural stress, σ_o (N/cm^2) is given by:

$$\sigma_o = M_x/Z \tag{3}$$

where $Z = I/c$ (cm^3) represents the section modulus of the beam. Results on such dependence will be illustrated in this paper using glass, graphite and boron fibers in epoxy matrix.

FORMULATION OF THE PROBLEM

We consider a beam under plane stress condition loaded by twin bending loads and simply supported as shown in Figure 1. The sides of the beam perpendicular to the loading direction have symmetrically placed fillets as shown in this figure.

Finite element analysis is employed to calculate the maximum stresses at the fillets. 4-node quadrilateral elements are used with (x and y direction) displacement functions:

$$U = \sum_{i=1}^{4} N_i u_i \quad , \quad V = \sum_{i=1}^{4} N_i v_i \tag{4}$$

$$X = \sum_{i=1}^{4} N_i x_i \quad , \quad Y = \sum_{i=1}^{4} N_i y_i \tag{5}$$

for this purpose.

In equations (3) and (4) N_i is an interpolation function (Bathe 1982), u_i and v_i indicate nodal displacements and x_i and y_i indicate nodal coordinates, respectively. The global (Cartesian) and local coordinates for a 4-node quadrilateral element are shown in Figure 2. Transformation between the global and local coordinates is done using the Jacobian operator:

$$\begin{Bmatrix} \dfrac{\partial}{\partial r} \\ \dfrac{\partial}{\partial s} \end{Bmatrix} = \begin{bmatrix} \dfrac{\partial x}{\partial r} & \dfrac{\partial y}{\partial r} \\ \dfrac{\partial x}{\partial s} & \dfrac{\partial y}{\partial s} \end{bmatrix} \begin{Bmatrix} \dfrac{\partial}{\partial x} \\ \dfrac{\partial}{\partial y} \end{Bmatrix} . \tag{6}$$

Hence, the Jacobian operator relates the local derivatives (such as $\partial/\partial r$) to the global derivatives (such as $\partial/\partial x$) and can be determined using equations (4) and (5). The derivatives $\partial u/\partial x$, $\partial u/\partial y$, $\partial v/\partial x$, and $\partial v/\partial y$ can, therefore, be determined to obtain strains using the strain-displacement relations.

Using the elasticity formulation for relating displacements to strains in matrix form:

$$\{\varepsilon\} = [B] \{u\} \tag{7}$$

the Jacobian is employed within the B matrix $\{\varepsilon\}^T = \{\varepsilon_x \ \varepsilon_y \ \gamma_{xy}\}$, and $\{u\}^T = \{u_1 \ v_1 \ u_2 \ v_2 \ u_3 \ v_3 \ u_4 \ v_4\}$.

The externally applied load ,P, is related to local stresses using the method of virtual work. Using the principal of energy balance between the external and internal forces we have:

$$\{P\} = \int [B]^T [Q^*] [B] \, dV\{u\} \tag{8}$$

where

$$[K] = \int [B]^T [Q^*] [B] \, dV \tag{9}$$

is the rigidity matrix and $[Q^*]$ is the transformed reduced stiffness matrix which will be defined in the next section. The differential volume expression dV of equations (8) and (9) can be written as

$$dV = \det J \, dr \, ds \tag{10}$$

in local coordinates.

The integral in equation (9) can be executed numerically. It is first written in the form

$$\int F(r,s) \, dr \, ds = \sum_{i,j} \alpha_{ij} F(r_i, s_j) \tag{11}$$

in two dimension. If we define

$$F = [B]^T [Q] [B] \det J, \tag{12}$$

then we can write

$$K = \int F \, dr \, ds \tag{13}$$

for the rigidity matrix.

Consequently,

$$[K] = \sum_{i,j} t_{ij} \, \alpha_{ij} \, F_{ij} \tag{14}$$

where F_{ij} is the F matrix for the nodes r_i, s_j, and t_{ij} is the thickness. In equation (14) $\alpha_{ij} = \alpha_i \, \alpha_j$ terms depend on the r_i, s_j values and are determined based on the Gauss Legendre numerical integration, (Bathe, Rao 1982). We note that α_{ij} values and the number and location for r_i, s_j affect the accuracy of the numerical integration which can be improved by increasing the number of nodes.

Once the rigidity matrix is obtained the displacements can be calculated by using

$$\{u\} = [K]^{-1} \{P\}. \tag{15}$$

Using the displacement values one can then calculate the stress vector $\{\sigma\}^T = \{\sigma_x \ \sigma_y \ \tau_{xy}\}$ using

$$\{\sigma\} = [Q^*] \{\varepsilon\} = [Q^*] [B] \{u\}. \tag{16}$$

Boundary Conditions

The boundary conditions for the filleted bar are shown in Figure 3. The finite element mesh used is also shown in Figure 3. It contains 880 elements and 945 nodes. For the unidirectional laminates the fiber direction was varied from 0° to 90° from the loading direction using 15° increments.

Analysis was performed for the whole plate in order to determine the stress concentration factors. For calculations, fillet radii was varied between 0.5 cm and 1.5 cm in 0.25 cm increments.

Material Properties

For the unidirectional lamina loaded at arbitrary angles to the principal material directions, the stress-strain relation is given by

$$\left\{\begin{array}{c} \sigma_x \\ \sigma_y \\ \tau_{xy} \end{array}\right\} = \left[\begin{array}{ccc} Q^*_{11} & Q^*_{12} & Q^*_{16} \\ Q^*_{12} & Q^*_{22} & Q^*_{26} \\ Q^*_{16} & Q^*_{26} & Q^*_{66} \end{array}\right] \left\{\begin{array}{c} \varepsilon_x \\ \varepsilon_y \\ \gamma_{xy} \end{array}\right\} \tag{17}$$

In the above equation the matrix Q_{ij} is the transformed reduced stiffness matrix with Q_{ij} elements defined by:

$$Q^*_{11} = Q_{11} C^4 + 2(Q_{12} + 2Q_{66}) S^2C^2 + Q_{22} S^4$$

$$Q^*_{12} = (Q_{11} + Q_{22} - 4Q_{66}) S^2 C^2 + Q_{12} (S^4 + C^4)$$

$$Q^*_{22} = Q_{11} S^4 + 2(Q_{12} + 2Q_{66}) S^2 C^2 + Q_{22} C^4 \tag{18}$$

$$Q^*_{16} = (Q_{11} - Q_{12} - 2Q_{66}) SC^3 + (Q_{12} - Q_{22} + 2Q_{66}) S^3 C$$

$$Q^*_{26} = (Q_{11} - Q_{12} - 2Q_{66}) S^3C + (Q_{12} - Q_{22} + 2Q_{66}) SC^3$$

$$Q^*_{66} = (Q_{11} + Q_{22} - 2Q_{12} - 2Q_{66}) S^2 C^2 + Q_{66} (S^4 + C^4)$$

where S and C are sine and cosine of the angle θ that the loading axis "x" makes with the principal material (fiber) direction "1".

In equation (18) the Q terms are defined by the following:

$$\begin{aligned} Q_{11} &= E_1 /(1 - \upsilon_{12} \upsilon_{21}) \\ Q_{12} &= (\upsilon_{12} E_2)/(1-\upsilon_{12} \upsilon_{21}) = (\upsilon_{21} E_1)/ (1-\upsilon_{12} \upsilon_{21}) \\ Q_{12} &= Q_{21} \\ Q_{22} &= E_2 /(1 - \upsilon_{12} \upsilon_{21}) \\ Q_{66} &= G_{12} \end{aligned} \tag{19}$$

where E_1 and E_2 are the composite's moduli of elasticity along the fiber (longitudinal) and transverse directions respectively, G_{12} is the shear modulus and $\upsilon_{12} = - \varepsilon_2/\varepsilon_1$ and, $\upsilon_{21} = - \varepsilon_1 / \varepsilon_2$ are the Poisson's ratios related by

$$\upsilon_{12} E_2 = \upsilon_{21} E_1. \tag{20}$$

The material properties E_1 , E_2 , G_{12} ,υ_{12} for the three types of composites: glass-epoxy, boron-epoxy and, graphite-epoxy used for analysis in this paper are shown in Table 1 (Jones, 1975).

RESULTS AND DISCUSSION

The variation of the stress concentration factor, K, with the fillet radius, r, is shown in Figures 4, 5, and 6 for glass-epoxy, boron-epoxy, and graphite-epoxy laminates, respectively. As observed for isotropic materials (Shigley and Mischke 1989) a general decrease is observed for stress concentration at the fillets when the fillet radius is increased.

The variation of the maximum normal stress, σ_{max}, with the fillet radius, r, is shown in Figures 7, 8, and 9 for glass-epoxy, boron-epoxy, and graphite-epoxy laminates, respectively. The general trend is an increase in the maximum stresses with increases in the fillet radius.

For all three types of fiber reinforcement (i.e. glass, boron, and graphite), the largest magnitudes for the stress concentration factor, K, and the maximum normal, σ_{max}, occur when the filleted bar is loaded at $\theta = 15°$ to the fiber reinforcement direction. The corresponding minimum values occur when $\theta = 90°$.

The descending order for the K and σ_{max} values from maximum to minimum correspond to graphite, boron and glass fiber reinforcements respectively for fiber reinforcement angles of up to and including 45°. This trend of increases in the stress concentration factors when graphite fibers are used instead of boron fibers or when boron fibers are used instead of glass fibers was also observed by A. Turgut et. al (1993) for composite plates containing u-notches.

For fiber reinforcement angles of 60°, and 75° the of graphite fibers still result in highest K and σ_{max} values but boron fibers replace glass fibers to obtain minimum K and σ_{max} magnitudes.

With 90° fiber reinforcement the difference between the K and σ_{max} values obtained with the three different fibers is considered insignificant.

These results are summarized in Table 2.

CONCLUSIONS

The results of our analysis confirmed the expected dependence of the stress concentration factor on the geometry of a discontinuity, which, for our case, is a fillet on a rectangular bar in bending. Furthermore, our results illustrated the dependence of the stress concentration factor, K, on the type of fibers used in a composite bar and also to their orientation with respect to the applied load. Graphite, boron and glass fibers, were used to illustrate the dependence of K values on the fiber type.

The use of graphite fibers resulted in highest magnitudes in stress concentration factors as well as in maximum normal stress values at the fillet locations. The highest increase in the stress concentration factors occured when the glass fibers were replaced by graphite fibers. This increase has been calculated to be as high as 86% for the largest fillet radius considered and with fiber reinforcement angle of 15°. For the smallest fillet radius considered, the corresponding increase in K was 81% at 15° fiber reinforcement angle.

REFERENCES

Sancaktar, E., 1990, in "Adhesives and Sealants, Testing and Analysis: Static and Dynamic Fatigue Testing", Engineered Materials Handbook, Volume 3: Adhesive and Sealants, ASM International, pp. 349-372.

Agarwal, B.D. and Broutman, L.J., 1990, Analysis and Performance of Fiber Composites, Wiley Interscience, New York.

Bathe, K.J., 1982, Finite Element Procedures in Engineering Analysis, Prentice Hall, Englewood Cliffs, NJ.

Rao, S.S., 1982, The Finite Element Method in Engineering, Pergamon Press, New York, NY.

Jones, R..M., 1975, Mechanics of Composite Materials, Hemisphere Publishing Company, New York, NY.

Shigley, J.E., and Mischke, C.R., 1989, Mechanical Engineering Design 5th ed., McGraw-Hill Book Co. New York, NY.

Turgut, A., Arslan, N., and Sancaktar, E., 1993, in Reliability, Stress Analysis, and Failure Prevention - 1993 ASME DE-Vol. 55, R.J. Schaller (ed.) pp. 125-134.

Material	E_1 (N/cm^2)	E_2 (N/cm^2)	G_{12}(N/cm^2)	υ_{12}
Glass-Epoxy	5.38 x 10^6	1.8 x 10^6	0.89 x 10^6	0.25
Boron-Epoxy	20.68 x 10^6	2.07 x 10^6	0.69 x 10^6	0.30
Graphite-Epoxy	20.68 x 10^6	0.52 x 10^6	0.26 x 10^6	0.25

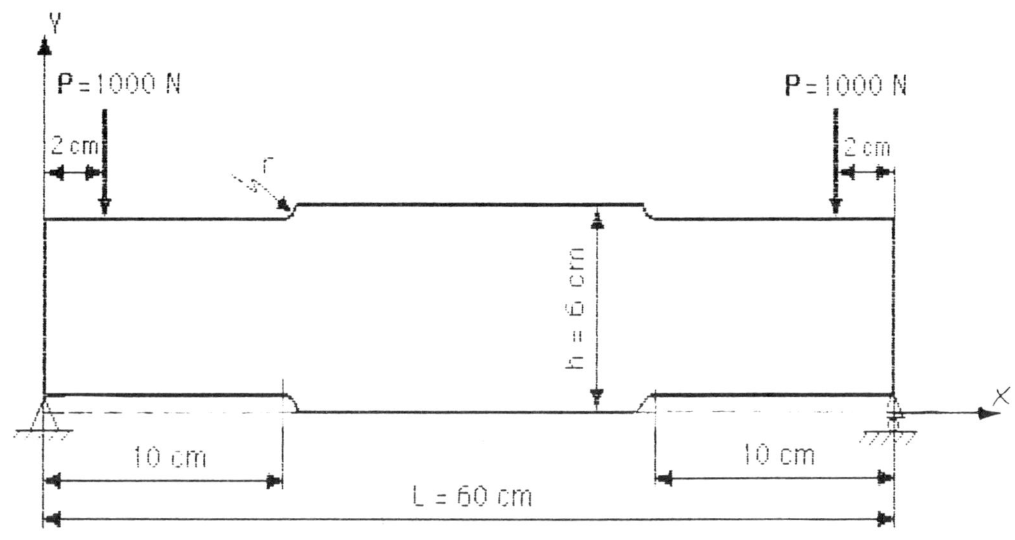

FIGURE 1. DIMENSIONS AND LOADING PATTERN FOR THE FILLETED BAR

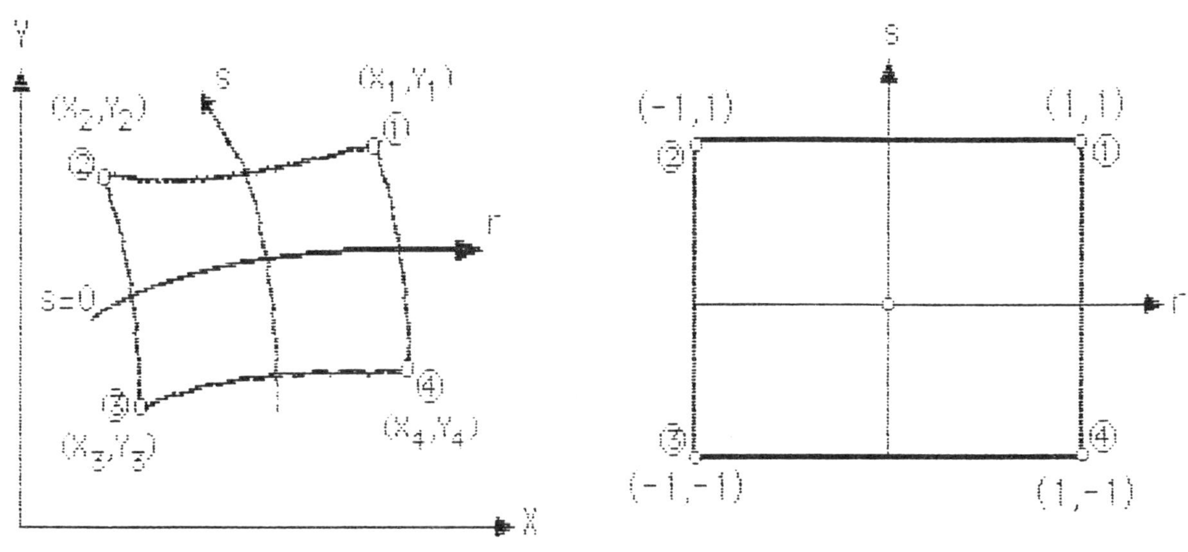

GLOBAL COORDINATES   LOCAL COORDINATES

FIGURE 2. 4-NODE QUADRILATERAL ELEMENT

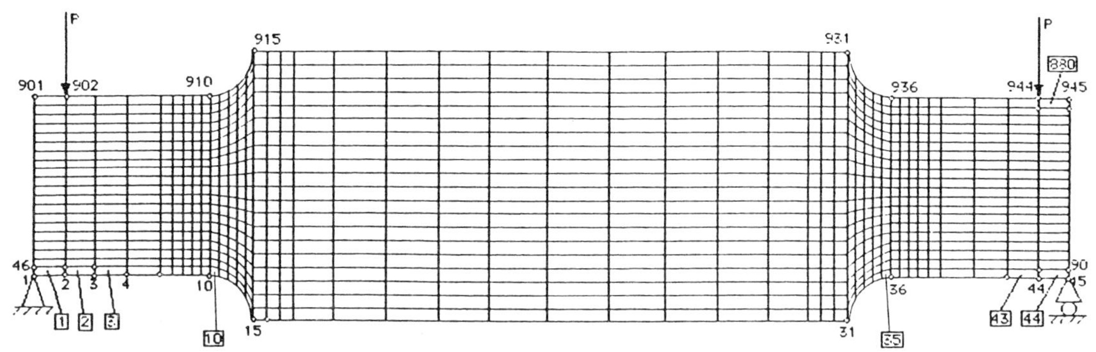

FIGURE 3. THE FINITE ELEMENT MESH USED

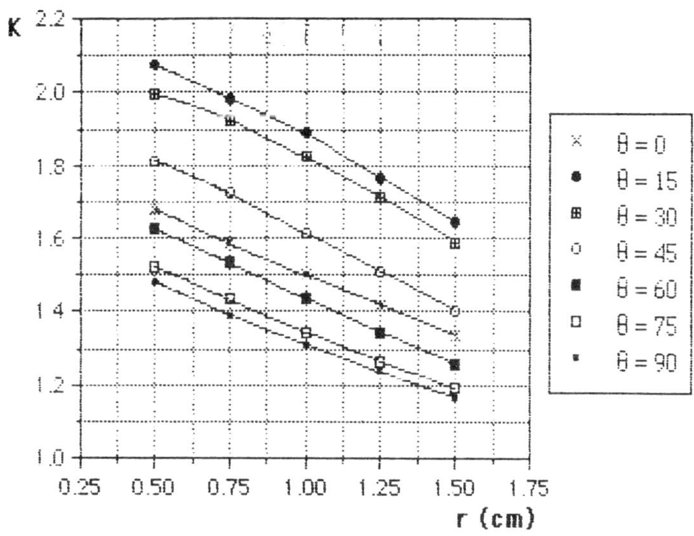

FIGURE 4. VARIATION OF THE STRESS CONCENTRATION FACTOR WITH FILLET RADIUS
AND FIBER ANGLE FOR GLASS-EPOXY COMPOSITE

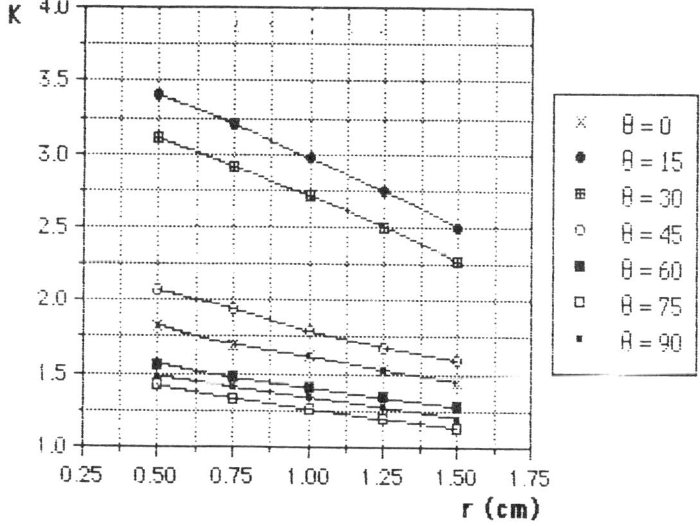

FIGURE 5. VARIATION OF THE STRESS CONCENTRATION FACTOR WITH FILLET RADIUS
AND FIBER ANGLE FOR BORON-EPOXY COMPOSITE

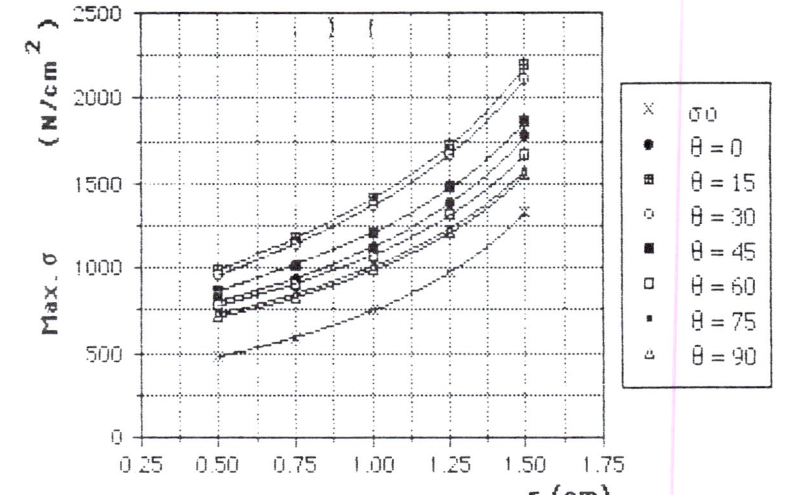

FIGURE 6. VARIATION OF THE STRESS CONCENTRATION FACTOR WITH FILLET RADIUS AND FIBER ANGLE FOR GRAPHITE-EPOXY COMPOSITE

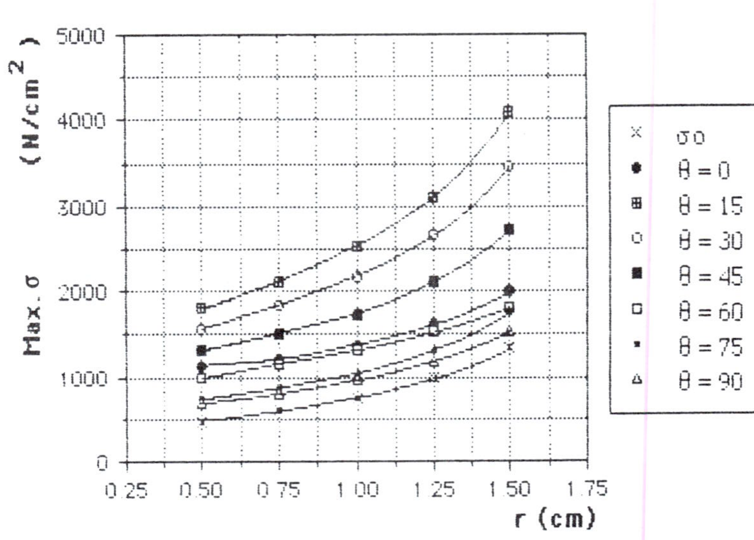

FIGURE 7. VARIATION OF THE MAXIMUM NORMAL STRESS WITH FILLET RADIUS AND FIBER ANGLE FOR GLASS-EPOXY COMPOSITE

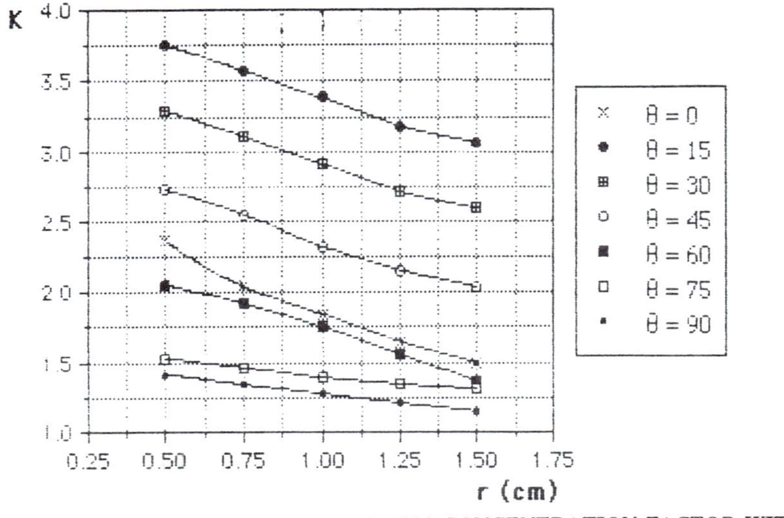

FIGURE 8. VARIATION OF THE MAXIMUM NORMAL STRESS WITH FILLET RADIUS AND FIBER ANGLE FOR BORON-EPOXY COMPOSITE

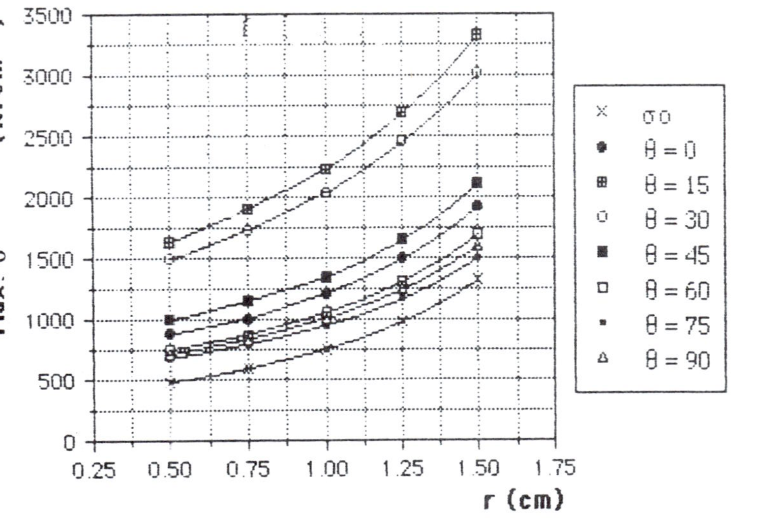

FIGURE 9. VARIATION OF THE MAXIMUM NORMAL STRESS WITH FILLET RADIUS AND FIBER ANGLE FOR GRAPHITE-EPOXY COMPOSITE

TABLE 2. THE VARIATION OF THE STRESS CONCENTRATION FACTOR AND THE MAXIMUM NORMAL STRESS WITH FIBER REINFORCEMENT ANGLE AND FILLET RADIUS

| Fillet Radius (cm) | FIBER REINFORCEMENT ANGLE, θ | | | | | | | | | | | | | |
| | STRESS CONCENTRATION FACTOR, K | | | | | | | MAXIMUM NORMAL STRESS, σ_{max} (N/cm^2) | | | | | | |
	0°	15°	30°	45°	60°	75°	90°	0°	15°	30°	45°	60°	75°	90°
0.50	1.681	2.073	1.997	1.813	1.625	1.523	1.481	806.880	995.040	958.718	870.496	780.255	731.263	711.276
0.75	1.585	1.985	1.925	1.723	1.532	1.432	1.392	939.259	1176.290	1140.930	1021.500	908.109	849.008	825.397
1.00	1.499	1.887	1.822	1.614	1.436	1.346	1.312	1124.430	1415.250	1367.060	1210.610	1077.010	1010.150	984.182
1.25	1.418	1.766	1.712	1.509	1.345	1.265	1.235	1389.720	1729.959	1677.620	1478.610	1318.060	1239.740	1210.560
1.50	1.335	1.646	1.590	1.402	1.257	1.188	1.164	1780.890	2194.666	2120.850	1869.780	1676.450	1584.880	1552.650
GLASS-EPOXY														
0.50	1.831	3.401	3.107	2.071	1.566	1.421	1.485	878.942	1632.48	1491.36	994.146	752.045	682.082	712.974
0.75	1.700	3.210	2.917	1.942	1.475	1.334	1.410	1007.730	1902.22	1729.92	1151.310	874.213	790.836	935.796
1.00	1.619	2.972	2.715	1.792	1.404	1.261	1.342	1214.260	2229.00	2036.35	1344.540	1053.450	946.397	1006.830
1.25	1.526	2.745	2.505	1.682	1.338	1.192	1.272	1495.040	2688.98	2454.04	1647.920	1311.350	1168.350	1246.060
1.50	1.446	2.501	2.261	1.592	1.273	1.130	1.204	1928.000	3334.66	3015.70	2122.950	1698.520	1507.240	1605.630
BORON-EPOXY														
0.50	2.365	3.755	3.290	2.735	2.057	1.535	1.418	1135.34	1802.41	1573.20	1312.80	987.361	736.800	681.293
0.75	2.047	3.577	3.112	2.557	1.925	1.465	1.345	1213.37	2119.71	1944.15	1515.26	1152.597	868.148	797.129
1.00	1.844	3.382	2.911	2.325	1.756	1.394	1.278	1383.45	2536.50	2183.25	1743.75	1317.000	1045.500	958.747
1.25	1.650	3.180	2.715	2.160	1.558	1.347	1.213	1616.64	3115.10	2659.59	2115.92	1526.204	1319.500	1188.630
1.50	1.498	3.065	2.596	2.041	1.362	1.311	1.151	1998.39	4086.66	3461.33	2721.33	1816.000	1748.530	1535.530
GRAPHITE EPOXY														

DE-Vol. 87, Reliability, Stress Analysis, and Failure Prevention
Issues in Emerging Technologies and Materials
ASME 1995

ON THE ROLE OF THE CHEMICAL POTENTIAL
IN PREDICTING LONG LIFE PERFORMANCE OF
POLYMERS AND POLYMERIC COMPOSITES

John F. Maguire and Peggy L. Talley
Materials and Structures Division
Southwest Research Institute
San Antonio, Texas

ABSTRACT

This paper presents new theoretical approaches for assessing the impact of environmental factors on the performance and design lifetime prediction of structures fabricated from polymers and polymeric composites. The method focuses on the role of the chemical potential in such problems and presents an analysis based on calculation of the chemical potential for systems which do not meet the usual criteria of the thermodynamic limit and which may exist in low dimensionality. The method has been shown to provide predictions which are in much better agreement with experiment than those which have been used heretofore and removes many of the "anomalous" diffusion dichotomies which continue to arise in the literature.

1.0 INTRODUCTION AND BACKGROUND

With the increased use of polymers and polymeric composites as structural materials in applications ranging from dental materials to aircraft engines, a need has been recognized to develop better methods to assess the impact of environmental, chemical, biological, and mechanical factors on the performance and design lifetime of such structures. Historically, the approach to this question has been largely heuristic and has not been well founded in chemical physics. For example, the work of Miller and Maguire (1995) has shown clearly that reliance on essentially parametric expressions such as the Arrhenius equation, which were developed for quite different purposes, can lead to errors of at least a factor of three in lifetime prediction. Detailed consideration of the microscopic mechanisms of defect transport in polymers (Bendler, 1995) has led to a much better understanding of the essential physics of defect transport in polymers and the role which such processes play in the ultimate failure of a component or structure.

Moreover, when chemical factors are operative, such as the presence of acidic environments or the presence of high humidity, structures such as helicopter rotor blades which had a design life of order 2000 hours have been withdrawn from service in a mere 20 hours. Clearly, if we can not predict lifetime to better than two orders of magnitude we can in no way have confidence in the current approaches to lifetime prediction. The essential difficulty is that the interaction between the polymer (or composite) and the environment takes place at an interface. This interface may be of thickness a few molecular diameters and so is technically "small" in the thermodynamic sense, i.e., the normal macroscopic laws of classical thermodynamics do not apply in this limit. Also, there will be a tendency for diffusion and perhaps strong physico-chemical interaction so that, by definition, the system is not in equilibrium. We have then a situation which requires a formalism which can handle the non-equilibrium thermodynamics (Prigogine, 1961) of a small system. Over the last twenty years, there have been very significant advances in our understanding of the basic physics of thermodynamically small systems, particularly those pertaining to interfaces (Rowlinson and Widom, 1982), and these advances offer the promise of much improved methodology in engineering.

To date, most studies of moisture diffusion in polymers and composites attempt to rationalize the experimental data in terms of Fickian diffusion. While a number of authors have recognized the inadequacy of the approach (Springer, 1988), there have been few attempts to make a systematic improvement. Indeed, rather than extend the theory to attempt to provide a unified treatment there has been a tendency to "classify" the behavior in terms which are essentially arbitrary. Essentially, if the mass uptake of solvent goes as $\sqrt{t}$ the process is defined as "Case 1", if it goes as t it is classified as "Case 2" while if it is somewhere in between it is deemed anomalous.

In developing a model for problems of this sort, it is usual to take Fick's second law as a starting point, i.e.,

$$\frac{\partial c}{\partial t} = D \nabla^2 C \qquad (1)$$

where c is the concentration, D is the mass diffusivity, and t is time. For the case of thermal diffusion, Eq. (1) is known to work very well. Exact analytic results are available for a number of geometries and boundary conditions. In one dimension the solution is well known (Crank, 1975)

$$c(x,t) - C_\infty \left\{ 1 + 2 \sum_{n-1}^{\infty} \cos\left[\frac{n\pi x}{L}\right] e^{\frac{-[n^2 \pi^2 D t]}{L^2}} \right\} \qquad (2)$$

where C_∞ maximum concentration and L in the distance from the surface to the center of the sample.

In those cases where closed form analytic results are not possible, numerical methods have been highly refined and will give solutions to essentially any required degree of accuracy. At the outset, it is necessary to recognize the essential physics underlying Eq. (1). The only assumption of importance in the derivation of Eq. (1) is that the flux is held to be directly proportional to the gradient. In some applications, such as heat flow, this is an excellent approximation. In others, such as material flow, it can be an exceedingly poor approximation. To see this, consider a "thought" experiment in which a fluid such as water is in contact with a polymer such as polyethylene. There is, of course, a very large gradient at the interface between the two materials. In this case, Eq. (1) makes the absurd prediction that the water will mix with the polyethylene in all proportion! Also, statements such as (Halse, et al., 1994), "The full derivation of this equation (2) is given by Parker et al. for the equivalent problem of the diffusion of a heat pulse into a thermally conducting sheet" illustrate clearly the widespread though mistaken belief that the physics of material transport is closely related to that of radiation transport. As we shall see, the question of diffusion of a fluid into a tortuous polymeric matrix is a problem of a fundamentally different sort.

The inadequacy of this approach is perhaps best illustrated in Figure 1. This figure shows a magnetic resonance image of the water concentration through the mid-section of glass-reinforced polycarbonate composite. The sample size was about 1 cm^3. It is abundantly clear from this image that the moisture profile is qualitatively very different from that which would be expected from Fickian diffusion. In particular, there is evidence that water transport along the fiber direction has been enhanced and also that there is a tendency for "puddle" formation. It is almost as if the water in the composite had a structure like that of water undergoing a spinodal decomposition. As we shall see, this is precisely what would be expected if the driving force for diffusion is treated in a proper fashion.

The remainder of this paper is organized as follows. Section 2 provides a discussion of the thermodynamics of small systems and develops a theory of fluid diffusion into polymers and composites. Section 3 presents some preliminary results using using model calculation and provides at least a partial resolution of some anomalous and conflicting results which have appeared in the recent literature.

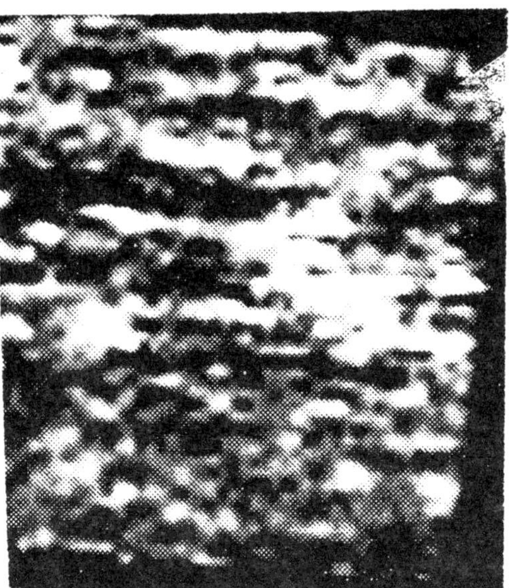

FIGURE 1. MAGNETIC RESONANCE IMAGE OF THE WATER CONCENTRATION IN THE MID-SECTION OF A GLASS-REINFORCED POLYCARBONATE COMPOSITE. THE FIBER DIRECTION RUNS FROM LEFT TO RIGHT. BRIGHT AREAS DENOTE HIGH WATER CONCENTRATION.

2.0 THE MODEL

Our starting point is the recognition that in a diffusion problem of this sort the fundamental physical laws will be those that pertain to small systems. In the usual elementary treatment of thermodynamics and statistical mechanics, the basic idea is that one is calculating averages for systems that are very large and which are not subject to an applied field. In reality, of course, no system is infinitely large and even very large systems are confined by vessels in which the interfacial region with the wall of the container can be considered small. For example, if p is the pressure, and T is the temperature, a schematic of the projections of the equation of state is usually depicted as shown in Figure 2(a). Only if the system is infinitely large will the liquid-gas and solid-liquid coexistence regions be perfectly horizontal and sharp. There are three lines of first order transitions which intersect in a triple point and the liquid-gas curve terminates in a bicritical point. Figure 2(b) shows a schematic of the chemical potential as a function of temperature (or any field variable) as a phase transition is traversed. If we take the solid liquid transition as an example, then if $\mu > \mu_T$, the system will be completely solid and if $\mu < \mu_T$, it will be completely fluid. However, as pointed out by Rowlinson and Swinton (1982), if $\mu = \mu_T$, exactly, the system can be completely solid, completely liquid, or any arbitrary mixture of both. We have then the apparent paradox that a large system is homogeneous even if it contains more than one phase. Such considerations have, interestingly enough, been largely ignored in the past by computer simulators. The reasons for this are well-known results of fluctuation theory. In a large system in the macrocononical ensemble, the fluctuation in the number of particles is proportional to $\sqrt{N}$ in each of the pure phases so as N gets large, the relative fluctuation becomes minuscule, $\sim 1/\sqrt{N}$. When $\mu = \mu_T$, the fluctuations in N become of order N and are thus never small. In such situations the fluctuations in the local thermodynamic properties can completely dominate the behavior of the system. At the solid-fluid transition, for example, the system could be all solid, all fluid, or any mixture in between. At this point, measurement of the size and range of the density fluctuations at and near a surface will provide very detailed information on the kinetics of the growth of the interface.

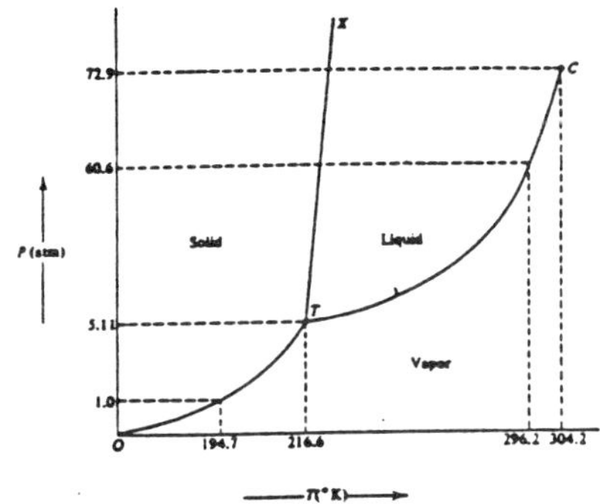

(a) PT projection of equation of state

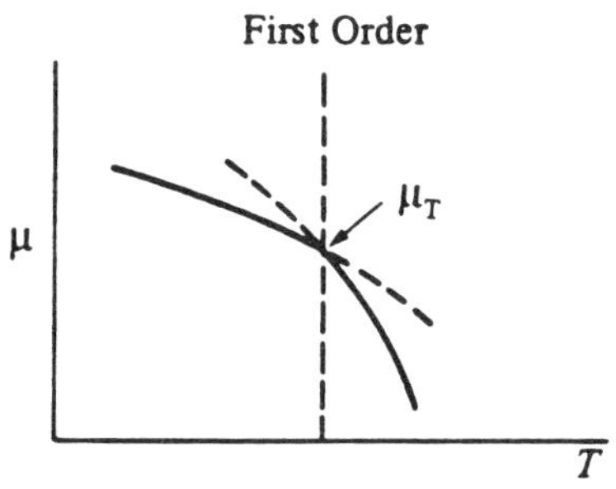

(b) Chemical potential as a function of temperature

FIGURE 2.

In order to produce the non-uniform system of our everyday experience, one must apply an external field of which the very weak gravitational and surface fields are the most ubiquitous, i.e.,

$$\mu = \mu(r) + w(r) + v(r) \qquad (3)$$

and

$$w(r) = mg(h - h_o) \qquad (4)$$

where m is the mass of a single molecule, g is the acceleration due to gravity, and h is the height above some reference height h_o and v(r) is the surface potential. For example, for the Lennard-Jones solid the surface potential is given by

$$v(r) = 8\pi\epsilon\rho\left[\left(\frac{\sigma}{r}\right)^9 - \left(\frac{\sigma}{r}\right)^3\right] \qquad (5)$$

where r the distance above the surface (which usually defines the reference height). The surface potential provides a perturbation which is longer range (r^{-3}) than the intermolecular interaction (r^{-6}). At the point of transition it is the gradient in this surface term plus that of the gravitational term which is responsible for the development of mass fluxes within the interfacial region. It is this gradient which is responsible for the structure and morphology of the interface. The behavior of a fiber-reinforced composite, which is made up almost completely of such interfaces, is thus dominated by such effects.

2.1 The Point Approximation

The surface term provides a sizable perturbation to the fluid, and the gravitational term, w(r), is also important. Note that if w(r) or v(r) have spatial **gradients**, even minute gradients, then the system will have two or more phases; a solid phase, a surface phase, and a fluid phase in equilibrium. Moreover, we might, even on the basis of these simple arguments, infer that the magnitude and form of this applied gradient may have an effect, possibly a substantial effect, on the morphology and structure of the phases which are formed. Notice that in the above case the system is, strictly speaking, a small system in that the length scale is determined by $\dfrac{dw(r)}{dr} \sim mg$. For a liquid at the triple point, this "gravitational" length $\left(\dfrac{kT}{mg}\right)$ is about 10 km. For a liquid system, the pair radial distribution function g(r) decays to unity on a length scale of order 2.0 nm. Under these circumstances, it is essentially exact to assume that the chemical potential is a functional of the local density and write

$$\mu(r) = \mu[\rho(r), T] \qquad (6)$$

Equation (6) states that the chemical potential and therefore all of the physical properties of the system are equal to those in the field-free state in a fluid of the appropriate density and temperature. It can be used, therefore, to calculate the chemical potential in bulk phases.

The important point for our purposes is to recognize that there are two common situations in which Equation (6) is **not** valid. The first is near a critical gas-liquid point or consolute point in a liquid mixture. Here, the correlation length is no longer determined by the range of the intermolecular potential, but diverges to an infinite length. The other is at a solid-fluid surface. In this case, the external field is a combination of both gravity and the surface potential. In such circumstances, the external field varies on the same distance scale as the oscillation in the radial distribution function so that here too, especially at the point of transition, minor variation in the field may have a huge effect on structure. It is for this reason that gravity and surface forces play such an important role in moisture diffusion. In order to explain interfacial phenomena and growth, it is necessary then to extend the argument to take into account the gradient effects. There are two levels of approximation at which this may be attempted.

2.2 The Local Approximation

The chemical potential, $\mu(\mathbf{r})$, and the free energy density, $\phi(\mathbf{r})$, have formally exact definitions that are not useful except for the purpose of conducting an evaluation using molecular dynamics or Monte Carlo integration (Henderson, 1992). The crudest approximation is the point approximation of Equation (6), which will hold only when the inhomogeneity is weak on the scale of the correlation length.

Following earlier attempts by Laplace and Rayleigh, van der Waals (Rowlinson and Widom, 1982) recognized that it is a free energy that is required to describe an interface, and developed his celebrated "square gradient" approximation for the free energy density,

$$\phi(\mathbf{r}) = \phi[\rho(\mathbf{r}), T] + \frac{1}{2} m \, |\nabla \rho(\mathbf{r})|^2 + \dots \tag{7}$$

where m is given by

$$m = -\frac{1}{6} \int r_{12}^2 \, U_{att}(r_{12}) \, dr_{12} \tag{8}$$

and U_{att} is the attractive long range part of the intermolecular potential.

For many years, the significance of this result was not appreciated until it was independently derived in slightly different form by Cahn and Hilliard (1958). The Cahn-Hilliard equation has been widely used to describe small systems such as wetting phenomena, by Rowlinson and Widom (1982). The need to extend the treatment beyond the point approximation for crystal growth has been well illustrated in the recent theoretical investigations of Kupferman et al. (1995) on the coexistence of symmetric and parity-broken dendrites in a channel. These authors have solved the diffusion equation for crystal growth (Eq. 12) using the assumption that μ can be approximated by the Gibbs-Thompsom LOCAL equilibrium relation. Interestingly enough, they find that for a large range of parameters stable symmetric and stable parity broken solutions co-exist. The branches of parity-broken solutions have their origin in symmetric solutions and arise from standard bifurcations.

2.3 Non-Local Approximation

The recent monograph edited by Henderson (1992) presents an excellent treatment of the non local approximation within the context of "non homogeneous fluids." The free energy density is regarded as a functional expansion of the formally exact expression of the free energy density.

$$\phi(\mathbf{r}_1) = \phi[\rho(\mathbf{r}_1)] - \frac{1}{2} kT \rho(\mathbf{r}_1) \int c(\mathbf{r}_1, \mathbf{r}_2) [\rho(\mathbf{r}_2) - \rho(\mathbf{r}_1)] \, d\mathbf{r}_2 \tag{9}$$

where $c(\mathbf{r}_1, \mathbf{r}_2)$ is the direct correlation function between points 1 and 2, which is related to the total correlation function

$$h(\mathbf{r}_1, \mathbf{r}_2) = g(\mathbf{r}_1, \mathbf{r}_2) - 1 \tag{10}$$

where $g(\mathbf{r}_1, \mathbf{r}_2)$ is the two body distribution function, given by the Ornstein-Zennike equation

$$h(\mathbf{r}_1, \mathbf{r}_2) = c(\mathbf{r}_1, \mathbf{r}_2) + \int c(\mathbf{r}_1, \mathbf{r}_3) \rho(\mathbf{r}_3) h(\mathbf{r}_3, \mathbf{r}_2) \, d\mathbf{r}_3 \tag{11}$$

The distinction between these approximations is not one of merely increasing numerical accuracy. We have already discussed the inadequacy of the point approximation (though even it is much better than Fickian diffusion), but these changes will lead to quite significant changes in the molecular systems with attractive tails that fall off as $1/r^m$.

The non-local approximation contains correlation functions such as $c(\mathbf{r}_1, \mathbf{r}_2)$, which we do not know much about. Note that in Equation (13) the density, which now depends on position, is inside the integral (sometimes called the OZ2 equation) and that this identity requires the use of closure relations, such as the hypernetted chain approximation (HNCA), the mean spherical approximation (MSA), or the Percus-Yevick approximation.

We may, therefore, write a generalized Smoluchowski-Langevin equation,

$$\frac{\partial}{\partial t} c(r, t) = D \nabla^2 \mu + \delta \qquad (12)$$

$$\mu = R T \ln c / c_0 \qquad (13)$$

where δ represents a fluctuation term. In this equation, the chemical potential μ is given by Equation (6) when we are far from an interface. This will be sufficient for our purposes here, though we should bear in mind that if we wish to model the fiber-matrix interface and hope to reproduce the observed granularity, then the free energy densities [Equations (7) and (9)] will have to be used. Clearly, in order to test the various theories for $\mu(r)$, it will be essential to have data in which the gravitational term $w(r)$ can be turned off and the consequences examined in terms of the structure and morphology of the system. The presence of both surface and gravitational terms in Equations (6) and (12) implies that in a small system such as the interfacial zone the gravitational perturbation may greatly affect the surface and near surface spectrum of density fluctuations. Notice that Eq. (14) does not include a convective term. If such an effect were present, it will, of course, be quite impossible to sort out the relative importance of the above contributions.

3.0 RESULTS AND DISCUSSION

Section 2 has provided a systematic overview of how the problem of fluid diffusion into polymers and composites is treated at three levels of approximation. Clearly the "granularity" which is clearly evident in the concentration field shown in Figure 1 is a real effect and is not caused by noise in the MRI image. This is likely to be true for all polymers since there will be inhomogeniety in the local concentration, cross-linked density etc. In order to model instabilities of this sort, it is likely that these more sophisticated theoretical approaches will be required. This will be exceedingly important, for as seen in Figure 1, the local concentrations in small volume elements may be an order of magnitude greater than what might be expected in terms of the overall bulk average. From an operational point of view, application of the full theory offers a number of significant computational challenges. Extending the theory to the level of the point approximation presents a relatively minor incremental difficulty in that all that is required is that the concentration field be mapped on to the chemical potential, i.e., an algebraic transformation is required. This approach is explored below. The local approximation requires computation of the gradient under the Laplacian while the non-local approximation will require the computation of a convolution integral at each step in the finite difference discretization. This will require very significant computational resources.

However, at the level of the point approximation, deviation from Fickian behavior can de treated in a straightforward way. The essential point is to recognize that the free energy of mixing is a concave function. Also, when the polymer is saturated with moisture, the chemical potential of a water molecule in the interior of the plastic is equal to the potential in pure water. If we denote the saturated concentration as c_∞ and the moisture content of the dry material as c_0 we have,

where μ is the chemical potential, R is the gas constant, T is the absolute temperature, and c_∞ is the concentration at saturation, which, for present purposes we may define as a reference state with zero chemical potential. Expanding Eq. (13) around the initial moisture content we may write,

$$\mu = R T \ln \left(\frac{c_0}{c_\infty} \right) + \sum_{n=1}^{n_\infty} (-1)^{n+1} \frac{(c - c_0)^n}{n \, c_0^{\, n}} \qquad (14)$$

Notice that the first order term (n=1) in Eq. (14) corresponds to Fickian diffusion while the higher order terms take into account progressively more non-Fickian behavior. Figure 3 shows fluid uptake curves for both processes. A concentration field, obtained by substituting Eq. (13) into Eq. (12) and neglecting fluctuations, is shown in Figure 4. First, it is seen that qualitatively the Fickian and non-Fickian weight gain curves are very different. It is important to bear in mind that the diffusion coefficient is the same for both curves. The inclusion of a chemical potential contribution has the following qualitative effect. At low absorption, the presence of the chemical potential term acts in the same fashion as increasing the concentration gradient. This has the effect of causing more moisture absorption than would be expected on the basis of the Fickian model. This has been observed in the careful experiments of Mohlin (1988). As the absorption progress towards saturation the uptake becomes slower than would be predicted on the basis of Fickian diffusion. It is easy to see how confusion might occur and systems could get "classified" into groups rather than analyzed in terms of a universal function.

The weight-gain curves have been plotted in the usual root fashion. However, note that one must go to very short times when taking the initial slopes if gross error in estimation of the diffusion coefficient is to be avoided. The essential point is that the driving force for diffusion should be properly taken into account, for otherwise, an equation such as (12), cannot be used to extract transport coefficients. In the present work we have assumed that the kinetics of transport can be related to gradients in thermodynamic potentials and have not addressed the important question of how thermodynamic and kinetic potentials may differ.

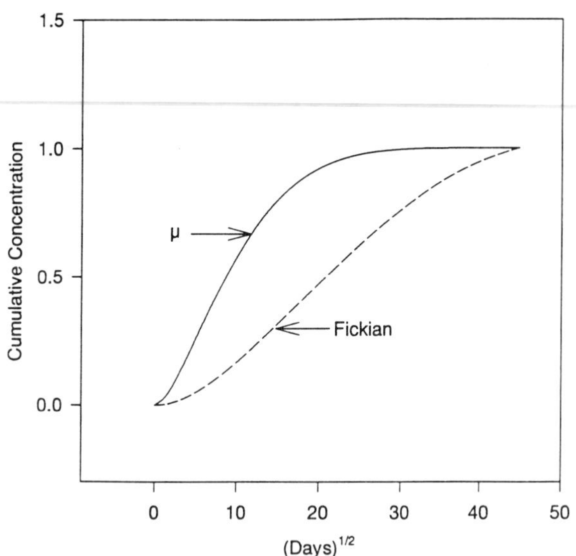

FIGURE 3. CUMULATIVE CONCENTRATION FOR FICKIAN AND POINT DIFFUSION MODELS. NOTE THAT THE CHEMICAL POTENTIAL DRIVES INITIAL DIFFUSION MUCH MORE RAPIDLY THAN FICKIAN. ALSO AVERAGE SHORT-TIME SLOPES DIFFER BY ABOUT A FACTOR OF FIVE.

LIST OF REFERENCES

Bendler, J. T., 1995, *ASME Conference Proceedings,* San Francisco, California.

Crank, J., 1975, *The Mathematics of Diffusion,* 2nd edition, Clarendon Press, Oxford.

Gao, P., and Mackley, M. R., 1994, *The Royal Society Proceedings; Mathematical and Physical Sciences,* Vol. 44, No. 1921, p. 207.

Halse, M. R., Rahman, H. J., and Strange, J. H., 1994, *Physica B,* Vol. 203, pp. 169-185.

Henderson, D., editor, 1992, *Fundamentals of Inhomogenous Fluids,* Marcel Dekker, New York.

Kupferman, R., Kessler, D. A., and Jacob, E. B., 1995, *Physica,* Vol. A213, pp. 451-464.

Miller, M. A., and Maguire, J. F., 1995, *ASME Symposium Proceedings,* San Francisco, California.

Mohlin, T., 1988, *Environmental Effects on Composite Materials,* Vol. 3, Chapter 13, p. 163, Technomic Publishing Co., Lancaster, Basel.

Prigogine, I., 1961, *Introduction to Thermodynamics of Irreversible Processes,* Wiley, New York.

Rowlinson, J. S., and Swinton, F. L, 1982, *Liquids and Liquid Mixtures,* 3rd edition, Butterworths, London.

Springer, G. S., 1988, *Environmental Effects on Composite Materials,* Vol. 3, Technomic Publishing Co., Lancaster, Basel.

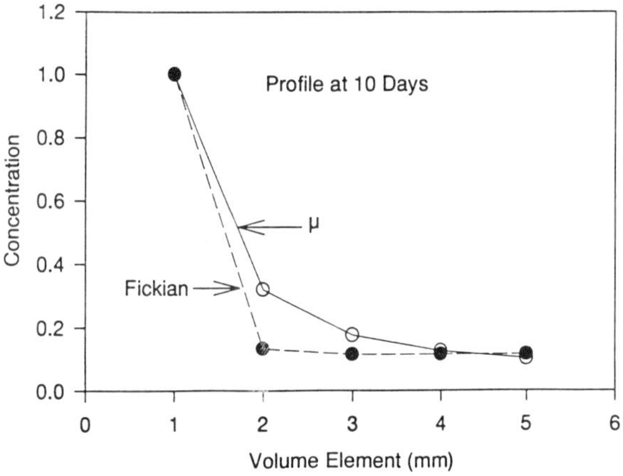

FIGURE 4. CONCENTRATION PROFILE AFTER 10 DAYS. NOTE THAT THE CURVES CROSS SO THAT LEAST SQUARES FITTING CAN GIVE DIFFUSION COEFFICIENTS WHICH ARISE FROM A CANCELLATION OF ERRORS.

A CUMULANT EXPANSION APPROACH TO
PREDICT THE SERVICE LIFETIME OF POLYMERS
AND POLYMER COMPOSITES

Michael A. Miller and John F. Maguire
Materials and Structures Division
Southwest Research Institute
San Antonio, Texas

ABSTRACT

The lifetime prediction of polymers and polymer composites has traditionally been approached by constructing an Arrhenius plot of the measured rate constants, as determined from mechanical failure experiments over an elevated temperature range, and extrapolating the best-fit line to low temperatures. This approach is fundamentally inaccurate for polymeric systems because kinetic rate constants cannot be accurately extrapolated through phase transitions that may occur at elevated temperatures, and the data are seldom linear at lower temperatures which are relevant to predicting service lifetime. In fact, the empirical results for even the simplest chemical systems are shown to deviate from linearity when extrapolated to low temperatures, which generally indicates that the rate constants being measured are a combination of rate constants for elementary processes with different activation energies. Therefore, the method of linear extrapolation of kinetic rate constants can not be expected to provide a meaningful prediction of service lifetime for complex systems such as polymers and polymer composites. The present work demonstrates an approach by which a cumulant expansion is used to determine the mean, variance, and higher moments of the probability distribution of activation energies that are inherent to complex polymeric systems. The coefficients of the cumulant expansion function are determined subsequent to non-linear regression analysis of the temperature dependence of the rate constants. This method provides a direct and more accurate means of equating the desired service temperature of a polymeric system to its lifetime prediction.

1.0   INTRODUCTION

The prediction of lifetime or time to failure (tf) of polymeric materials has traditionally been approached by assuming that the rate constants obey the Arrhenius equation $\ln k = \ln A - (Ea/RT)$ over an exceedingly broad temperature range, where quantity Ea is the activation energy (in Joules per mole) and A is the frequency factor or pre-exponential factor. The usual procedure reported in the literature is to make a few measurements of the time to failure for induced stresses at elevated temperatures, where the relaxations processes are of such length so as to make the measurement times convenient, then extrapolate these measurements linearly using the Arrhenius equation to much lower temperatures; the actual temperature at which the material will be in service. Such extrapolations have been performed even when the experimental data clearly deviates from linearity at lower temperatures, in such cases many workers have either ignored the deviation or have attributed this phenomenon to poor experimental data. The latter point is naive when one considers that temperature and time can be routinely measured with millikelvin and micorsecond resolution, respectively.

The notion that the kinetics of mechanical failure follow the Arrhenius equation is non-physical. It is essential to recognize that over a broad temperature range, the temperature dependence of the rate constants are not adequately represented by this formalism and additional terms are needed. For example, in chemical kinetics, it has long been recognized (Westenberg and de Haas, 1967; Shultz and LeRoy, 1964) that deviations from the Arrhenius law indicate that the rate constant being measured is a combination of rate constants for a number of elementary processes with different, and indeed, a distribution of activation energies. However, with regard to lifetime predictions of mechanical failure, little or nothing has been done to adequately represent this behavior or adopt this concept.

The aim of the present work is to introduce an alternative method for predicting the lifetime of polymeric materials which takes into full account any non-linear behavior of the experimental measurements over a broad temperature range. The method presently described is derived from purely statistical concepts and is therefore useful in determining the probability for time to failure of a material under stress at any desired service temperature. The premise of this approach is that the non-linear behavior associated with the temperature dependence of the rate constants is governed by a statistical distribution of activation energies for the failure mode. That is, the activation energies are temperature dependent over a distribution, and, for the purpose of this discussion, such energies are considered to be normally distributed. The important point is that this approach is in accord with physical reality and does not force the actual behavior of the failure process into linear terms.

The mathematical basis for this treatment is that the Arrhenius equation can be viewed to be a special case of a cumulant expansion: $\ln\phi(\tau) = \Sigma\kappa_n(i\tau)^n/n!$, where $\phi(\tau)$ is the characteristic function of a stochastic variable, and κ_n is the nth cumulant, with n taken from zero to infinity. In stochastic theory, the first and second cumulants, κ_1 and κ_2, are the mean (μ) and variance (σ^2), respectively, while higher order cumulants are combinations of these moments. Thus, in view of the form of the Arrhenius equation, where only κ_0 and κ_1 are relevant, the first cumulant is proportional to the mean energy of activation for failure and the constant κ_0 is the frequency factor. The variance κ_2 vanishes for this special case as it must if the rate process can be described by only one energy of activation.

More generally, one can easily see from this approach that cumulants of higher order, in particular the variance κ_2, give rise to the non-linear behavior of the expansion function which is needed to adequately describe rate processes involving a multitude of transitions with their respective energy of activation.

In this work, the cumulant expansion approach is applied to time to failure data for polyethylene as a function of applied stress at different temperatures. The relationship between time to failure and applied stress suggested by Wolstenholme and Stark (Wolstenholme and Stark, 1964), and Coleman and Knox (Coleman and Knox, 1957) is modified in terms of the said treatment such that no assumptions are made as to how the induced stress varies with the time to failure. Non-linear regression analysis is used to fit the cumulant expansion function, taken to the second order, to the experimental stress and temperature data.

METHODS

The time to failure in polyethylene depends upon the operating temperature and the induced stress. The rate equation for the induced stress has been described by Wolstenholme and Stark (Wolstenholme and Stark, 1964), and Coleman and Knox (Coleman and Knox, 1957) to be related to the apparent free energy of activation for failure according to the following equation

$$\ln t_f(S,T) = \ln C - \ln T + (\Delta F/RT) + bS/T \qquad (1)$$

where S is the induced stress, T is the temperature, ΔF is the free energy of activation for failure, and $\ln(C)$ and b are constants. Equation (1) can be written in the form

$$\ln t_f(S;T) = A + bS/T \qquad (2)$$

where now A is the intercept of the isothermal stress data. That is, at zero induced stress $\ln t_f = \ln C - \ln T + (\Delta F/RT)$. Because no assumptions will be made as to the actual behavior of the time to

failure versus stress data for each isotherm, equation (2) is expanded as a second order cumulant function to derive the zero stress intercepts (A). The form of this expansion is

$$\ln t_f(S;T) = \kappa_0 + \kappa_1(iS) + 1/2\kappa_2(iS)^2 \qquad (3)$$

where now κ_0 is the intercept of the non-linear expansion and κ_1 contains b/T. In practical calculations the factor i is cumbersome, and since S is a set of real numbers, it may be avoided by setting $iS = \sigma$ such that the second order cumulant expansion for each isotherm at temperature T becomes

$$\ln t_f(\sigma;T) = K_0 + K_1(\sigma) + K_2(\sigma)^2. \qquad (4)$$

Non-linear least squares regression fits of equation (4) to the isothermal time to failure versus induced stress data permits a more exact determination of the intercepts $K_0 = A$. From equation (1), the set of real intercepts K_0 added to $\ln T$ allows the determination of the apparent free energy of activation

$$\ln t_f(T;S=0) = \ln C + (\Delta F/RT). \qquad (5)$$

Here as before, a second order cumulant expansion of equation (5) allows the actual behavior of the time to failure (*i.e.*, $K_0 + \ln T$) versus T^{-1} data to be adequately described in non-linear terms. The cumulant expansion of equation (5) is given by

$$\ln t_f(T;S=0) = K'_0 + K'_1(T)^{-1} + K'_2(T)^{-2} \qquad (6)$$

where now, by means of a non-linear least squares regression fit of the $\ln t_f$ versus T^{-1} data, K'_1 is determined to be proportional to the mean apparent free energy of activation for the failure with variance (σ^2) equal to K'_2 .

RESULTS

Figure 1 depicts literature data (Brown and Lu, 1988) for the time to failure in a polyethylene single edge notched tensile specimen as a function of the applied stress at different temperatures (25, 42, 50, 60, 70, and 80 C). The isothermal data as presented in Figure 1 are overlaid with the best-fit line governed by equation (2) to illustrate how the experimental data deviates from an assumed linear relationship between time to failure and induced stress. These deviations are quite significant at the higher temperature isotherms, and would ordinarily yield erroneous values for the intercepts if one were to assume that equation (2) is valid.

The time to failure versus T^{-1} data are plotted in Figure 2 for both linear (Equation 2) and non-linear (Equation 4) extrapolation of the zero stress intercepts. The curves drawn through the points are the results of fitting the cumulant expansion function, equation (6), to the derived intercept data. The best-fit parameters for these equations are compared in Table 1.

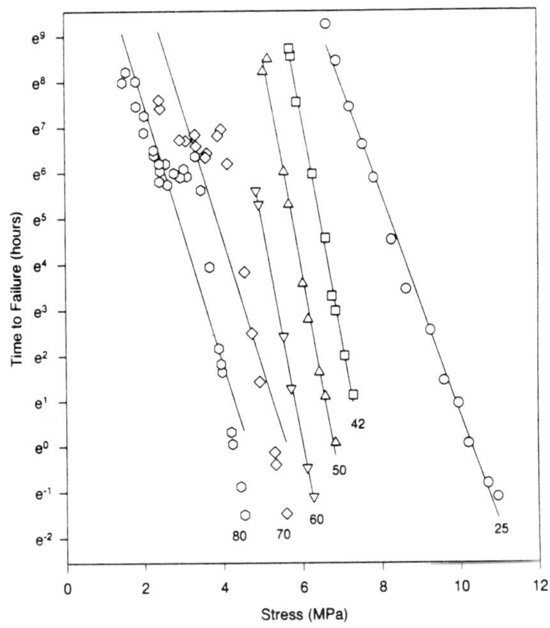

FIGURE 1. Time to Failure Versus Induced Stress Curves for Polyethylene at Various Isotherms.

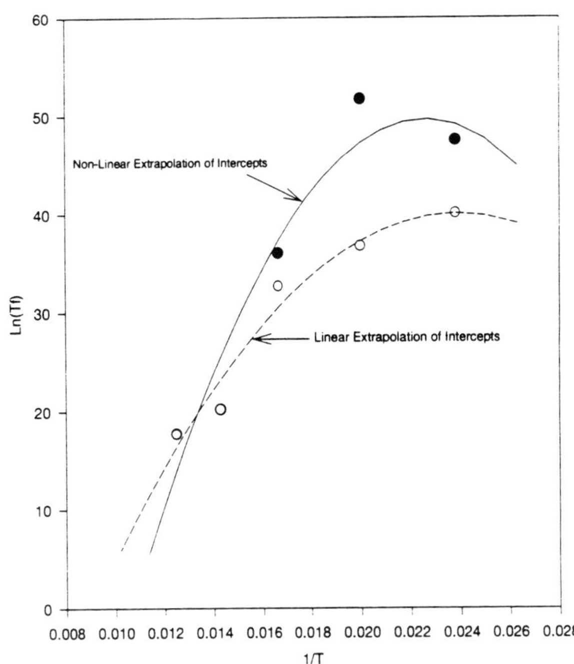

FIGURE 2. Time to Failure Versus Temperature (T^{-1}) Profiles Showing the Best-Fit Cumulant Expansion Curves Through the Extrapolated Intercepts.

TABLE 1. Non-Linear Regression Analysis Parameter Fits of Cumulant Expansion.

Parameter	Best-Fit Parameter Values	
	Equation (6) Using Linear Extrapolation of Intercepts (Eq. 2)	Equation (6) Using Non-Linear Extrapolation of Intercepts (Eq. 4)
K_0	-64.0	-129
K_1	8703	$1.58e^4$
K_2	$-3.64e^5$	$-3.49e^5$

Figure 2 clearly demonstrates significant curvature to the relationship between time to failure and temperature, which from a stochastic viewpoint suggests that the activation energy for this failure mode is not singular, but rather has a distribution of values with a significant variance which is dependent on temperature. In fact, the degree of curvature, which is predicated by the second cumulant or variance in the expansion, tends to suggest that the time to failure is bimodal at lower temperatures. Regardless of what long-term behavior may be inferred from these results, the important point is that this approach at least offers the opportunity to predict the service lifetime of the material at any given temperature in terms of probabilities using the mean and variance derived from the best fit second order cumulant expansion equation.

CONCLUSIONS

The cumulant expansion approach introduced in this work can adequately describe the non-linear temperature dependence of failure rate-processes for polymeric materials in terms of statistically meaningful parameters. The method is completely general and precise, and does not presume that failure rate data is governed by an Arrhenius relation. Application of the method to previously published time to failure data for polyethylene as a function of applied stress at several isotherms demonstrated that the activation energy for failure is temperature dependent, and that the Arrhenius relation is not, in general, valid. Moreover, use of the Arrhhenius equation in situations where pronounced cumulant of the semi-log plot is clearly apparent leads to gross errors in the predicted lifetime of components.

The cumulant method illustrated here leads to much more accurate lifetime prediction.

LIST OF REFERENCES

Westenberg, A. A., and de Haas, N., 1967, *J. Chem. Phys.* Vol. 47, p. 1393.

Schulz, W. R., and LeRoy, D. J., 1964, *Can. J. Chem.*, Vol. 42, p. 2480.

Wolstenholme, W. E., and Stark, C. F., 1964, Report No. 1, Contract DA-18-035-AMC-139(A).

Coleman, B. D., and Knox, A. G., 1957, *Textile Res. J.*, Vol. 27, p. 393.

Brown, N., and Lu, X., 1988, *Proceedings of the International Conference of Plastic Pipes VII.*

DE-Vol. 87, Reliability, Stress Analysis, and Failure Prevention
Issues in Emerging Technologies and Materials
ASME 1995

MAINTAINING A CONSISTENT CONFIGURATION IN A CONSTRAINT-BASED MECHANICAL DESIGN SYSTEM

A. Vincent Huffaker
Department of Manufacturing Engineering
Boston University
Boston, Massachusetts

Oded Z. Maimon
Department of Manufacturing Engineering
Boston University
Boston, Massachusetts

ABSTRACT

We consider here the class of mechanical designs that can be expressed as systems of nonlinear equations that describe the dimensions of the feature(s) being created. Systems of nonlinear equations typically have multiple possible solutions even though the designer usually has only one in mind. Iterative solution techniques used to solve systems of equations (e.g., Newton-Raphson) fall short due to their lack of control over which solution the system converges to. We introduce here an interval-based technique to efficiently solve this problem.

Variational design is a powerful paradigm for design, but has been limited in its application by the inability to characterize multiple solutions resulting from systems of nonlinear equations. Previous research efforts in variational design have introduced numerous methods for increasing the stability of iterative methods in the presence of multiple competing solutions. These methods can be categorized as either 1) requiring the user to manually monitor and correct the convergence process, or 2) requiring the user to define additional constraints or other information about the system.

Instead of these approaches, this paper introduces techniques for tracking a specific solution as the constraints are modified by the user. In this scenario, the user only has to help guide the search the first time the systems of constraints are solved. During subsequent modifications (as occur often in design), the techniques are able to automatically follow the trajectory of the solution as the parameters change so as to maintain a consistent configuration. The incremental design of a cantilever beam is used to demonstrate the solution trajectory tracking technique.

1. INTRODUCTION

We consider the class of mechanical designs that can be expressed as systems of nonlinear equations that describe the dimensions of the feature(s) being created. Systems of equations have typically been solved using iterative numerical techniques such as the Newton-Raphson method. When the equations are nonlinear, however, there are often multiple competing solutions to the system, but only one is usually acceptable to the designer. When there are multiple solutions, simple iterative techniques fall short due to their lack of control over which solution the system converges to. We develop here an interval based variational technique to efficiently solve this problem.

A *design* is defined in terms of parts and relations. Every part can be described by a set of *dimensions*. A *constraint* is a relation among the dimensions that must be satisfied. Finally, a *configuration* is a set of dimension values that fully describe a feasible part.

In many mechanical design problems, a part is described by a system of constraints. In variational design, the entire system of constraints for a part (often, nonlinear equations) is solved simultaneously. Each constraint is translated into a mathematical equation and the system of equations is then solved. This is in contrast to a parametric design system where the constraints are solved sequentially as they are entered. Variational design systems offer several advantages including the ability to solve problems whose dimensions cannot be calculated through a sequential series of equations (Mackrell, 1993).

Typically a system of nonlinear equations has many possible solutions (configurations). There has always been difficulty, however, in handling this situation. Historically, there have been two different approaches to this problem: either the user can manually select the correct solution (either by guiding the solving algorithm so as to wind up at the correct solution or by picking from list of possible solutions), or the user can over-constrain the system of equations so there is only one solution. Neither approach is satisfactory. The first requires too much extraneous user interaction during the design. When any of the external parameters are modified, the user should not have to manually guide the search. The second approach requires too much information from the user. It is hard enough for the user to fully-define any system of constraints. Requiring the user to enter even more constraints is not practical in general, and may also unnecessarily exclude other feasible solutions.

Instead of these approaches, this paper introduces the idea of following the trajectory of the desired solution as it moves as the constraints are modified. Then, a new method for following a trajectory is introduced that is based on rigorous sensitivity analysis using interval analysis techniques.

In our system, MECHAD, the specification of the design is done by either directly specifying the values of the dimensions or by specifying the values of higher-order attributes (e.g., volume, mass, maximum stress). The user inputs as many constraints as are necessary to fully constrain the part. The user only has to help guide the search the first time the systems of constraints are solved. During subsequent modifications (as occur often in design), MECHAD automatically follows the trajectory of the solution as the parameters change so as to maintain a consistent configuration.

The algorithm to reconverge to the same correct solution after a constraint modification utilizes two techniques from interval analysis. First, rigorous sensitivity analysis is used to create guaranteed enclosures of the dimension values as the attribute values are changed. Second, interval methods are used to find and verify the uniqueness of the solution within the range of the attribute values being changed. The values of the attributes are incrementally moved towards the final values as these two techniques are applied at each iteration to ensure a consistent reconvergence.

When applied to the case study presented in this paper, these techniques are shown to increase the stability of variational design. After modifying the given system of constraints, the Newton-Raphson method converged to the wrong solution. Using the iterative convergence method resulted in proper convergence if three or more iterations were used but incorrect convergence if one or two iterations were used. Conversely, MECHAD automatically reconverged to the correct solution with no additional user intervention (e.g., specifying the initial solution or the convergence rate).

1.1 Paper Outline

Section 2 discusses previous efforts in variational design and, in particular, the efforts they have made at handling the problem of multiple solutions. Section 3 develops the concept of solution trajectories and introduces the algorithm for maintaining a consistent solution, when constraints are modified. Section 4 demonstrates these techniques in the design of a cantilever beam with a uniformly distributed load and compares the design process when using other methods. Finally, Section 5 makes summary conclusions and discusses recommendations for future work. Appendix A gives an overview of interval analysis; Appendix B shows the the constraint model used for the beam design; and Appendix C defines the design paradigm that MECHAD solves by introducing a formal evolutionary model of design and describing MECHAD as an instance of the theory.

1.2 Nomenclature

The following nomenclature is throughout this paper to define the problem and the algorithms. In Appendix C, additional syntax will be introduced in order to formally define MECHAD's design paradigm in the same syntax of the design model introduced in (Maimon and Braha, 1995).

$\underline{b},\underline{q}$	Real number vectors used to denote the right hand side of the systems of equations (the attribute values)
f,Q	Real, continuous function, $\mathfrak{R}^n \mapsto \mathfrak{R}$ or $I\mathfrak{R}^n \mapsto I\mathfrak{R}$
$\underline{f},\underline{Q}$	Systems of equations, $\mathfrak{R}^n \mapsto \mathfrak{R}^n$ or $I\mathfrak{R}^n \mapsto I\mathfrak{R}^n$
$I\mathfrak{R}$	Set of real intervals
x,d	Real numbers used to denote either an unknown in general (x) or an unknown dimension (d)
$\underline{x},\underline{x}^i$	Vectors of real numbers of dimension n
$\underline{x}_j,\underline{x}^i_j$	Element j of vector $\underline{x}$ or $\underline{x}^i$
$[x]$	Interval x
$\mathrm{rad}([x])$	Radius of $[x]$
$\mathrm{mid}([x])$	Midpoint of $[x]$
$[a,b]$	$[x]$ with $\uparrow[x] = a$ and $\downarrow[x] = b$
$\alpha,\beta,\delta,\lambda$	Real numbers used in algorithms to modify the convergence

2. PREVIOUS WORK

A system of m independent equations describes the values of n unknown variables. A system of equations is characterized as follows:

under-constrained:	$m < n$
fully-constrained:	$m = n$
over-constrained:	$m > n$

In addition, the term *uniquely-constrained* describes a system of equations that describes exactly one solution.

Although a fully constrained parameter-dependent system of nonlinear equations usually has multiple solutions (not uniquely-constrained), typically the designer only has one in mind:

$$\underline{f}(\underline{x}) = \underline{b} \Rightarrow \underline{x} = \left\{ \underline{x}^1, \underline{x}^2, ..., \underline{x}^{i^*}, ..., \underline{x}^{P_1} \right\},$$

There is no general approach for deciding which of the solutions is the correct one - this is based on the designer's intent. In order to identify the correct solution, programs typically rely on the user to enter an initial guess, which is used as the initial solution for an iterative solution technique. When the designer decides to change the system of constraints by either choosing new constraints and/or defining new values of the constraints, not only does the desired solution change, but the entire set of possible solutions changes:

$$\underline{f}'(\underline{x}) = \underline{b}' \Rightarrow \underline{x} = \left\{ \underline{x}'^1, \underline{x}'^2, ..., \underline{x}'^{j^*}, ..., \underline{x}'^{P_2} \right\}.$$

Many prior research efforts in variational design have come across the problem of converging to the wrong solution. Seven different methods identified in the literature for maintaining a consistent solution are summarized below.

2.1 Initial Solution

Methods for solving a system of nonlinear equations, $\underline{f}(\underline{x}) = \underline{b}$, are all of the iterative form $\underline{x}^{k+1} = \underline{x}^k + \underline{\alpha}\underline{x}^k$ where $\underline{\alpha}$

is a corrective vector that is applied to the preceding iteration. Newton-Raphson, one of the most common methods, uses a correction vector based on the inverse of the Jacobian of the system of equations. Every method for solving systems of nonlinear equations depends upon an initial solution, $\underline{x}^0$, to control the specific solution it converges to. As such, every variational design program offers the user the ability to modify it (Beaty et al., 1994; Light, 1980; and Serrano, 1984).

2.2 Number Of Iteration Steps

If the values of the external parameters only change slightly, then the values of the current dimensions should be reasonably close to the new dimensions. Conversely, if the parameter values change drastically, then the current dimension values have almost nothing in common with the new values. Accordingly, this technique strives to maintain the same solution by reducing the size of the parameter changes.

When the user changes the values of any of the parameters, $\underline{f}(\underline{x}) = \underline{b}'$, they also can enter the number of iteration steps, m. The technique then iteratively steps towards the new value solving the system of equations,

$$\underline{f}(\underline{x}) = \underline{b} + \frac{(\underline{b}' - \underline{b})}{m} \cdot \beta, \ \beta = 1,...,m,$$

at each incremental value of the parameter, using the converged value at each iteration step, β, as the initial solution of the next iteration (Light, 1980 and Lin, 1981).

2.3 Relaxation Factor

Rather than requiring the user to specify the exact number of iteration steps to take for each parameter change, this method allows the user to simply slow down the convergence by multiplying the corrective vector by a scalar number (the relaxation factor): $\underline{x}^{k+1} = \underline{x}^k + \lambda \underline{\alpha x}^k$ where λ typically varies between 0 and 1 (Light, 1980; Lin 1981; and Gallaher, 1984).

2.4 Curve-Crawl Factor

Instead of multiplying the correction vector by a scalar number, this method uses a function of the magnitude of the correction vector of the previous iteration:

$$\underline{x}^{k+1} = \underline{x}^k + \lambda \left(\left| \underline{\alpha x}^{k-1} \right| \right) \underline{\alpha x}^k .$$

Ideally, this function would become close to 1 as the magnitude of the previous correction vector headed towards 0 (thereby not limiting a slow convergence) and would become smaller for larger magnitudes of the previous correction vector to slow down a faster convergence (Light, 1980 and Lin, 1981).

2.5 Feasibility Limits

The techniques mentioned so far rely solely on the user to figure out when the solver has converged to an unintended solution. Gallaher (1984) presents an alternative method in which feasibility limits are assigned to every variable. If the solver converges to a value outside the feasible range, then the relaxation factor is decreased (slowing the convergence down) and the convergence is repeated until either the variables remain inside their feasibility limits or a given number of attempts has been exhausted.

2.6 Manual Selection

All the previously mentioned techniques require the user to guide the convergence process itself. This method, in contrast, simply strives to find every solution to the system of equations and allows the user to select the desired solution. This approach is used in higher-level symbolic mathematical packages (e.g., Mathematica) and is stated as a future goal of Buchanan (1993). As all solutions are considered equal in worth, there is no attempt to automatically distinguish the correct solution from an unintended solution.

2.7 Over-Constraining The System

The final method is to have the user over-constrain the system until it is uniquely-constrained (Buchanan, 1993). As this may sometimes yield no solutions, both Borning (1992) and Wilson (1993) additionally give the user the option to prioritize the constraints (rather than requiring them all to be simultaneously satisfied).

2.8 Summary

The techniques discussed either require the user to interact with the solver, guiding its convergence towards the correct solution, or to input so much extra information that there is only one solution and, hence, no ambiguity. None of the techniques are satisfactory for our system. When any of the external parameters are modified, the user should not have to manually guide the search. Also, it is sufficiently challenging for the user to fully-define any system of constraints; requiring the user to over-constrain the system so that there is only one solution for all possible constraint values is not practical in general.

3. AN ALGORITHM FOR MAINTAINING A CONSISTENT CONFIGURATION

The problem of inconsistent configurations arises in variational design because anytime a system of constraints is changed, it is typically completely re-solved as if it is a new problem. In this section, we introduce the notion of the trajectory of a solution. Then, rigorous sensitivity analysis techniques are derived and integrated into a technique for following the trajectory of a given solution.

3.1 The Idea Of Solution Trajectories

In order to demonstrate solution trajectories, let us consider the very simple case of creating a point at the intersection of two circles and then modifying the radii of the circles. This problem has two unknowns (the x and y coordinates of the point) and two equations, so it is fully-constrained. However, there are typically two solutions (so it is typically not uniquely-constrained).

The constraints requiring the point to lie on each of the two circles form a system of two nonlinear equations:

$$
\underline{f}(\underline{d}) = \left\{ \begin{array}{l} \sqrt{\left(\underline{d}_1 - c_{x_1}\right)^2 + \left(\underline{d}_2 - c_{y_1}\right)^2} = \underline{b}_1 \\[2mm] \sqrt{\left(\underline{d}_1 - c_{x_2}\right)^2 + \left(\underline{d}_2 - c_{y_2}\right)^2} = \underline{b}_2 \end{array} \right.
$$

Setting $c_{x_1} = 0$, $c_{y_1} = 0$, $c_{x_2} = 3$, $c_{y_2} = 4$, $\underline{b}_1 = 4$, and $\underline{b}_2 = 2$ yields $\underline{d}$ = {1.00411, 3.87192} and {3.43589, 2.04808} (case I). Changing the radii to $\underline{b}_1 = 3$ and $\underline{b}_2 = 6$ changes the two possible solutions to $\underline{d}$ = {-2.51466, 1.636} and {2.27466, -1.956} (case II). Figure 1 demonstrates the scenario.

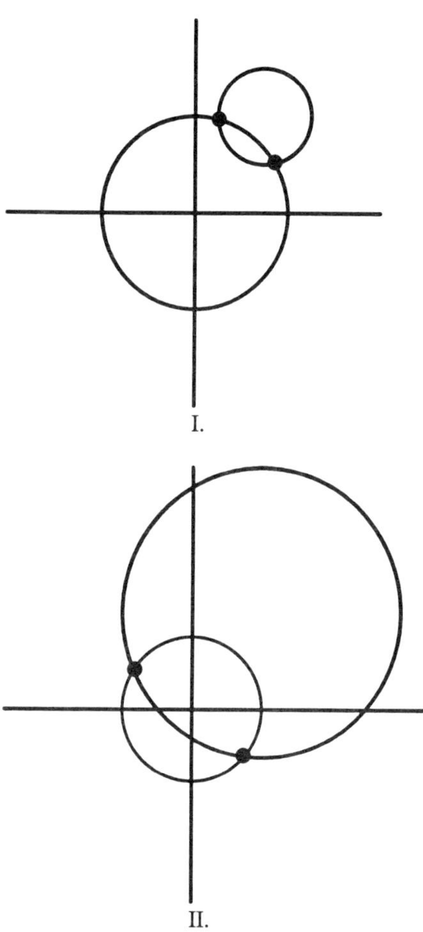

I.

II.

FIGURE 1: SIMPLE VARIATIONAL DESIGN EXAMPLE

If the original and new sets of parameter values are considered as two discrete events, the only way to maintain consistency of the design is to be able to distinguish individual solutions and be able to recognize which solution the user desires whenever the parameters are changed. While this is certainly possible for a simple geometric design problem, it is not possible in general for a system of nonlinear equations.

Rather than treating parameter changes as discrete events, we consider the continuous trajectory the solutions follow as the parameters are changed from the original values to the new values. Rather than trying to characterize the solutions,

MECHAD follows a specific solution as it moves along its solution trajectory. For instance, as the radii of the circles are changed, the two points each create a trajectory from their original position to their new position (as shown in Figure 2).

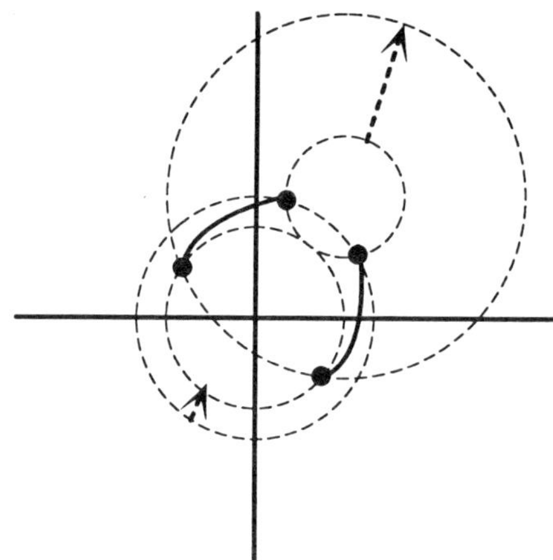

FIGURE 2: SOLUTION TRAJECTORIES

3.2 Linear Sensitivity Analysis

The problem of following a trajectory is often approached by linearizing the system (using its Jacobian) and using the simplified system of equations to approximate the new value as given by the mean-value theorem:

$$
\underline{f}(\underline{x}) - \underline{f}(\underline{y}) = \underline{f}'(\underline{z})(\underline{x} - \underline{y}) \text{ where } \underline{z} \in [\underline{x}, \underline{y}].
$$

As $\underline{z}$ is unknown, usually the originating point is used, resulting in the truncated Taylor series approximation:

$$
\underline{f}(\underline{x}) \approx \underline{f}(\underline{y}) + \underline{f}'(\underline{y})(\underline{x} - \underline{y})
$$

which can be manipulated as follows to find the estimate of the new dimension values:

$$
\underline{f}(\underline{x}) - \underline{f}(\underline{y}) \approx \underline{f}'(\underline{y})(\underline{x} - \underline{y})
$$

$$
\underline{f}'(\underline{y})^{-1} \cdot \left(\underline{f}(\underline{x}) - \underline{f}(\underline{y})\right) \approx (\underline{x} - \underline{y})
$$

$$
\underline{x}^e = \underline{y} + \underline{f}'(\underline{y})^{-1} \cdot \left(\underline{f}(\underline{x}) - \underline{f}(\underline{y})\right)
$$

After calculating the estimate, $\underline{x}^e$, the true value is calculated, $\underline{x}^*$. If the difference, $\left|\underline{x}^e - \underline{x}^*\right|$, is "close enough", the solution is assumed to have correctly followed the trajectory. If not, then $\underline{f}(\underline{x})$ is iteratively brought closer to $\underline{f}(\underline{y})$ until the estimate is good enough.

In the case of the point at the intersection of the two circles, the estimates and actual values of the new solutions are:

$\underline{y}$	$\underline{x}^e$	$\underline{x}^*$
{1.00411, 3.87192}	{-3.00453, 3.87841}	{-2.51466, 1.636}
{3.43589, 2.04808}	{4.56454, -1.7984}	{2.27466, -1.956}

The estimates may or not be reasonable approximations of the actual solutions. The problem with this approach is that the best selection for $\underline{f}(\underline{x})$ and the criteria for judging the acceptability of an estimate are not deterministic quantities. They are, themselves, guesses, subject to the effects of higher-order nonlinearities in the system of equations.

3.3 Rigorous Sensitivity Analysis

Rather than working with fixed floating-point representations of numbers, MECHAD uses interval analysis techniques to create guaranteed bounds for enclosures of a solution to a system of equations subject to change (see Appendix B for a review of interval analysis techniques). MECHAD solves the system of equations, $\underline{f}(\underline{x}) = \underline{b}$ when $\underline{b}$ varies from $\underline{b}^0$ to $\underline{b}^1$ $\left(\underline{f}(\underline{x}^0) = \underline{b}^0 \to \underline{f}(\underline{x}^1) = \underline{b}^1 \right)$.

The interval extension of the mean-value theorem (Krawczyk and Neumaier, 1985) guarantees the existence of an interval vector, $[\underline{z}]$, that exhibits the following property:

$$\underline{f}(\underline{x}^1) - \underline{f}(\underline{x}^0) \in \underline{f}'([\underline{z}])(\underline{x}^1 - \underline{x}^0) \tag{1}$$

where $\underline{x}^0, \underline{x}^1 \in [\underline{z}]$.

Substituting for $\underline{f}(\underline{x}^1)$ and $\underline{f}(\underline{x}^0)$ yields:

$$\underline{b}^1 - \underline{b}^0 \in \underline{f}'([\underline{z}])(\underline{x}^1 - \underline{x}^0) \tag{2}$$

Which can be represented as a linear interval system of the form $\underline{A}\underline{x} = \underline{b}$ where:

$$\underline{A} = \underline{f}'([\underline{z}]), \quad \underline{x} = \underline{x}^1 - \underline{x}^0, \text{ and } \underline{b} = \underline{b}^1 - \underline{b}^0 \tag{3}$$

The solution to this system, $\underline{x} = \underline{A}^H \underline{b}$ (defined as the hull of the solution set of the linear interval system), is enclosed by $\underline{x} = \underline{A}^{-1}\underline{b}$, where $\underline{A}^{-1}$ is the interval extension of the matrix inverse of $\underline{A}$ (which is computable if $\underline{A}$ is regular).

$$\underline{x} = \underline{x}^1 - \underline{x}^0 \supseteq \underline{A}^{-1}\underline{b} \tag{4}$$

$$\underline{x}^1 \supseteq \underline{x}^0 + \underline{A}^{-1}\underline{b} \tag{5}$$

$$\underline{x}^1 \supseteq \underline{x}^0 + \underline{f}'([\underline{z}])^{-1}(\underline{b}^1 - \underline{b}^0) \tag{6}$$

Equation (6) gives an initial guaranteed enclosure for $\underline{x}^1$. The bounds can be further tightened using the iterative solution techniques presented in Appendix A (e.g., Gauss-Seidel iteration).

3.4 Algorithm For Consistent Configuration During Reconvergence

Using equation (6) and techniques from (Hammer et al., 1993), (Neumaier, 1989), and (Neumaier, 1990), we have created an algorithm that uses interval analysis techniques to deterministically follow a solution trajectory. The central idea is to first iteratively expand $[\underline{z}]$ to be as large as possible (there must be a unique solution to $\underline{f}(\underline{x}^0) = \underline{b}^0$ in $[\underline{z}]$, and $\underline{f}'([\underline{z}])$ must be regular). Then, equation (6) shows how to compute guaranteed enclosures of estimates within $[\underline{z}]$. For each iteration, the estimate of $[\underline{x}']$ is extended as far as possible (it cannot extend outside of $[\underline{z}]$ and there must be a unique solution within each enclosure).

I. Initialize

$\quad \underline{b}' = \underline{b}^0$

$\quad \underline{x}' = \underline{x}^0$

$\quad \beta = 0.0$

$\quad \delta = 0.05$

II. Calculate $[\underline{z}]$

$\quad \alpha = 1.1$

$\quad mid([\underline{z}]) = \underline{x}'$ {Initialize $[\underline{z}]$ }

$\quad rad([\underline{z}]) = \underline{x}' / 10$

$\quad$ Solve $\underline{f}(\underline{x}) = \underline{b}'$ when $\underline{x} \in [\underline{z}]$

$\quad$ If there is a unique solution, $\tilde{\underline{x}} \in [\underline{z}]$, and $\underline{f}'(\underline{z})$ is regular then

$\qquad$ Repeat

$\qquad\quad$ {expand the search interval}

$\qquad\quad rad([\underline{z}]) = \alpha \cdot rad([\underline{z}])$

$\qquad\quad$ Solve $\underline{f}(\underline{x}) = \underline{b}'$ when $\underline{x} \in [\underline{z}]$

$\qquad$ Until there is no longer a unique solution, $\tilde{\underline{x}} \in [\underline{z}]$, or

$\qquad f'([\underline{z}])$ is not regular

$\qquad rad([\underline{z}]) = rad([\underline{z}]) / \alpha$

$\quad$ Else

$\qquad$ Repeat

$\qquad\quad$ {compress the search interval}

$\qquad\quad rad([\underline{z}]) = rad([\underline{z}]) / \alpha$

$\qquad\quad$ Solve $\underline{f}(\underline{x}) = \underline{b}'$ when $\underline{x} \in [\underline{z}]$

$\qquad\quad \alpha = \alpha / 1.1$

$\qquad$ Until there is a unique solution, $\tilde{\underline{x}} \in [\underline{z}']$, and $\underline{f}'([\underline{z}])$

$\qquad$ is regular

$\quad$ End If

III. Calculate guaranteed enclosure within $[\underline{z}]$

$\quad \beta = \beta + 0.1$

$\quad \tilde{\underline{b}} = \underline{b}'$

$\quad \underline{b}' = \underline{b}^0 + \beta(\underline{b}^1 - \underline{b}^0)$

$\quad [\underline{z}'] = \tilde{\underline{x}} + \underline{f}'([\underline{z}])^{-1}(\underline{b}' - \tilde{\underline{b}})$

$\quad$ Solve $\underline{f}(\underline{x}) = \underline{b}'$ when $\underline{x} \in [\underline{z}']$

$\quad$ Repeat

$\qquad$ {project the enclosure just past the boundary}

$\qquad$ If there is a unique solution $\tilde{\underline{x}} \in [\underline{z}']$ and $[\underline{z}']$ is within

$\qquad [\underline{z}]$ then

$$\beta = \min(1, \beta + 0.1)$$

$$\underline{b}' = \underline{b}^0 + \beta(\underline{b}^1 - \underline{b}^0)$$

$$[\underline{z}'] = \widetilde{\underline{x}} + \underline{f}'([\underline{z}])^{-1}(\underline{b}' - \widetilde{\underline{b}})$$

Until there is no longer an unique solution or $[\underline{z}']$ is no longer within $[\underline{z}]$

Repeat

    {compress the enclosure until valid}

$$\beta = \max(0, \beta \cdot (1 - \delta))$$

$$\underline{b}' = \underline{b}^0 + \beta(\underline{b}^1 - \underline{b}^0)$$

Solve $\underline{f}(\underline{x}) = \underline{b}'$ when $\underline{x} \in [\underline{z}']$

Until there is a unique solution and $[\underline{z}']$ is within $[\underline{z}]$

<u>IV. Iterate until</u> $\beta = 1$

    While $\beta < 1$

$$\underline{x}' = \widetilde{\underline{x}}$$

        Step II.

        Step III.

    End While

3.5 General Applicability Of The Algorithm

It may at first seem a limitation to limit the equations to be of the form $\underline{f}(\underline{x}) = \underline{b}$ (where the parameters are restricted to the right hand side) as opposed to more general formulations, $\underline{f}(\underline{x}, \underline{b}) = \underline{0}$ (where the parameters are allowed to be part of the equations themselves). It is claimed that the first form, however, can be used without loss of generality.

Consider the trajectory of a solution as it moves between the two systems of equations, $\underline{f}^0(\underline{x}^0, \underline{b}^0) = \underline{0}$ and $\underline{f}^1(\underline{x}^1, \underline{b}^1) = \underline{0}$. Plugging $\underline{x}^0$ into the second system of equations yields a typically nonzero value, $\underline{f}^1(\underline{x}^0, \underline{b}^1) = \underline{q}$. Since $\underline{b}^1$ is a constant vector, the trajectory can be seen to travel between the two systems of equations, $\underline{f}^1(\underline{x}^0) = \underline{q}$ and $\underline{f}^1(\underline{x}^1) = \underline{0}$, giving the desired formulation (parameters restricted to the right hand side).

4. DEMONSTRATION

In order to demonstrate these techniques, a cantilever beam is designed for a uniformly distributed load. For each kind of part to be designed in MECHAD, a constraint model is constructed defining the dimensions and all the higher-order constraints. The constraint model for the beam is given in Appendix B.

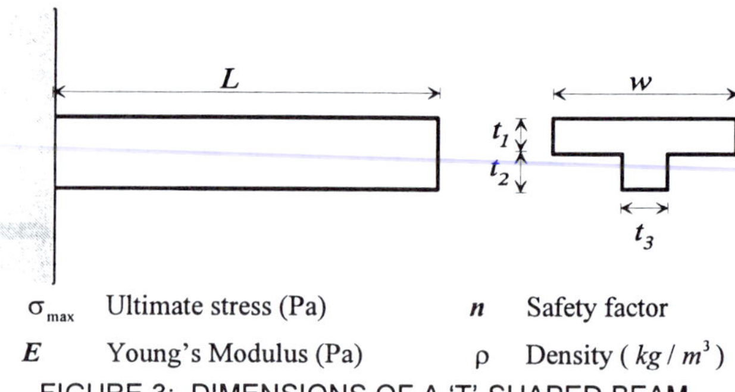

σ_{max}	Ultimate stress (Pa)	n	Safety factor
E	Young's Modulus (Pa)	ρ	Density (kg/m^3)

FIGURE 3: DIMENSIONS OF A 'T'-SHAPED BEAM

Additionally, the following attributes are defined for the beam in the constraint model:

Volume, V	Mass, M
Total Surface Area, S	Maximum Moment, M_{max}
Maximum Load, p_{max}	Maximum Allowable Load, p_{allow}
End Angle at p_{max}, θ_{max}	End Angle at p_{allow}, θ_{allow}
End Deflection at p_{max}, Δ_{max}	
End Deflection at p_{allow}, Δ_{allow}	

4.1 Brief Design Paradigm Description

In MECHAD, a design is represented as:

$$\underline{f}(\underline{d}) = \underline{b} \tag{7}$$

$$\underline{Q}(\underline{d}) \geq \underline{q} \tag{8}$$

Here, $\underline{f}(\underline{d})$ is the fully-constrained system of n equality constraints as entered by the user. The system is solved simultaneously to obtain the desired values of the dimensions, $\underline{d}$. $\underline{Q}(\underline{d})$ is the set of p qualitative specifications generated by the user to define the successful termination of the design process.

The design process begins with the set of qualitative specifications describing the successful termination of the design process (e.g., "produce for less than a given cost" and/or "have a maximum load greater than some value"), and a set of constraints that, when solved, produces a design that may or may not satisfy them. If not, the user iteratively modifies the constraints or their values until the design is satisfactory.

4.2 Design Execution

At the beginning of the design process, the designer has in mind a rough vision of what the part should look like and a initial set of qualitative design specifications (Table 1).

A first guess is then made at a constraining set of dimensions and attributes that will satisfy $\underline{Q}(\underline{d})$ (Table 2):

$\underline{Q}(\underline{d})$		$\underline{q}$
p_{allow}		2000 (kg)
l		0.20 (m)
-(M)	$\geq$	-(1.5) (kg)
$-(\Delta_{allow})$		-(0.0001) (m)
w		0.02 (m)

TABLE 1: QUALITATIVE DESIGN SPECIFICATION

$\underline{f}^0(\underline{d}^0)$		$\underline{b}^0$
t_1		0.01 (m)
t_2		0.015 (m)
t_3		0.01 (m)
p_{allow}		2000 (kg)
Δ_{allow}	$=$	0.0001 (m)
σ_{max}		$385 \cdot 10^6$ (Pa)
ρ		7850 (kg/m^3)
n		2.5
E		$200000 \cdot 10^6$

TABLE 2: FIRST DESIGN ITERATION

All but two of the part's dimensions (w and l) are directly constrained. Mathematically, there are 4 possible values of the free dimensions that a normal iterative constraint solver could converge to:

$$w = 0.01158303 \quad l = 0.18290615$$
$$w = 0.01158303 \quad l = -0.18290615$$
$$w = -0.26639935 \quad l = 0.144491157$$
$$w = -0.26639935 \quad l = -0.144491157$$

Of the four solutions, only the first one is possible so that is the one intended:

$$w = 0.01158303 \quad l = 0.18290615$$

Calculating the results of the qualitative specifications gives:

$p_{allow} = 2000$ (kg)	(acceptable)
$l = 0.18290615$ (m)	(unacceptable)
$\Delta_{allow} = 0.0001$ (m)	(acceptable)
$M = 0.038168$ (kg)	(acceptable)
$w = 0.01158303$ (m)	(unacceptable)

Since $\underline{Q}(\underline{d})$ is not satisfied, the design process continues.

Since w and l were both too small, the constraints are changed to increase their values. Additionally, the mass is increased to ensure there is enough material to allow dimensions of sufficient size. Table 3 gives the resulting system of constraints:

$\underline{f}^1(\underline{d}^1)$		$\underline{b}^1$
t_1		0.01 (m)
w		0.04 (m)
l		0.20 (m)
M	$=$	1.3 (kg)
Δ_{allow}		0.0001 (m)
σ_{max}		$385 \cdot 10^6$ (Pa)
ρ		7850 (kg/m^3)
n		2.5
E		$200000 \cdot 10^6$

TABLE 3: SECOND DESIGN ITERATION

Again, all but two dimensions (t_3 and t_2) are directly constrained. There are three possible solutions to this system of equations (including one impractical solution).

$$t_3 = -0.2504105 \quad t_2 = -0.0017092$$
$$t_3 = 0.0219435 \quad t_2 = 0.0195057$$
$$t_3 = 0.0450158 \quad t_2 = 0.0095083$$

There is no obvious way to simply deduce which solution is the correct one. While the first solution is obviously not the correct one since it is not practical, the other two solutions are both reasonable and under different circumstances, either one could be the desired solution. In order to compute which solution is the correct one, MECHAD creates and then follows the solution trajectory of the initial solution as the attribute values are changed.

To create the solution trajectory, the constraints are computed at the original dimension values, giving the values of the new constraints which yield the same original solution. The solution trajectory is then given by parametric curve:

$$\left\{ \underline{d} \mid \underline{f}^1(\underline{d}) = \underline{f}^1(\underline{d}^0) + t\left(\underline{f}^1(\underline{d}^1) - \underline{f}^1(\underline{d}^0)\right) \right\}, \; 0 \leq t \leq 1 \qquad (9)$$

While $\underline{d}^1$ is an unknown quantity, it is known that $\underline{f}^1(\underline{d}^1) = \underline{b}^1$:

$$\left\{ \underline{d} \mid \underline{f}^1(\underline{d}) = \underline{f}^1(\underline{d}^0) + t\left(\underline{b}^1 - \underline{f}^1(\underline{d}^0)\right) \right\}, \; 0 \leq t \leq 1 \qquad (10)$$

The solution trajectories for the three possible solutions are shown in Figure 4. By using the techniques from Section 4, MECHAD is able to follow the solution trajectory of the desired solution as it moves to its new, correct solution:

$$t_3 = 0.0219435 \qquad t_2 = 0.019505$$

The results of the qualitative specifications are as follows:

$p_{all} = 6181$ (kg)	(acceptable)
$l = 0.20$ (m)	(acceptable)
$\Delta_{allow} = 0.0001$ (m)	(acceptable)
$M = 1.3$ (kg)	(acceptable)
$w = 0.04$ (m)	(acceptable)

Since $\underline{Q}(\underline{d})$ is satisfied, the design process is completed.

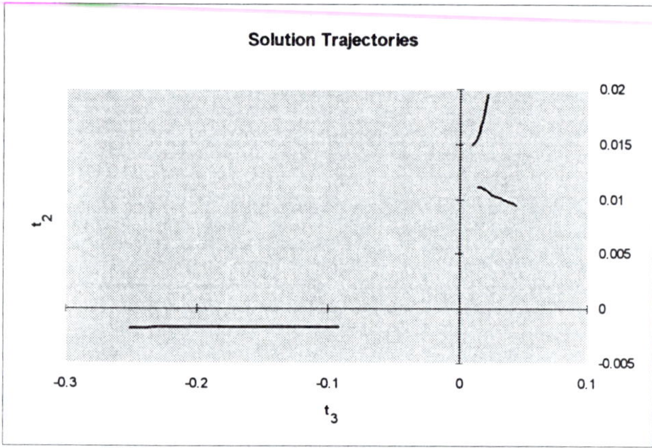

FIGURE 4: SOLUTION TRAJECTORIES

4.3 Comparison With Other Methods

For a comparison of other methods, we use Newton's method to solve the same system of equations, $\underline{f}^1(\underline{d}^1) = \underline{b}^1$ (using the FindRoot command in Mathematica). The initial value of each dimension is taken to be either the value directly constrained by the user or, if it does not exist, the value of the dimension from the previous design iteration. Table 4 shows the initial values of each dimension and the value that it converges to. As can be seen from the values of t_2 and t_3, Newton's method converges to the wrong solution.

Dimension	t_1	t_2	t_3
Initial Value	0.01	0.015	0.01
Converged Value	0.01	0.0095	0.04502
Dimension	w	l	n
Initial Value	0.04	0.2	2.5
Converged Value	0.04	0.2	2.5
Dimension	E	ρ	σ_{max}
Initial Value	$2 \cdot 10^{11}$	7850	$3.85 \cdot 10^8$
Converged Value	$2 \cdot 10^{11}$	7850	$3.85 \cdot 10^8$

TABLE 4: CONVERGENCE USING NEWTON-RAPHSON

In order to try to converge to the correct solution, iterative convergence (as reviewed in Section 2.2) is attempted and the number of iterations is increased until correct reconvergence is attained (Table 5).

	Dimension	Converged Value
Number Iterations = 2	t_1	0.01
	t_2	0.0095
	t_3	0.04502
	w	0.04
	l	0.2
	n	2.5
	E	$2 \cdot 10^{11}$
	ρ	7850
	σ_{max}	$3.85 \cdot 10^8$
Number Iterations = 3	t_1	0.01
	t_2	0.01951
	t_3	0.02194
	w	0.04
	l	0.2
	n	2.5
	E	$2 \cdot 10^{11}$
	ρ	7850
	σ_{max}	$3.85 \cdot 10^8$

TABLE 5: CONVERGENCE USING ITERATION STEPS

By increasing the number of iterations, Newton's method is able to converge to the correct solution. Finally, we tried reducing the speed of the convergence by modifying the relaxation factor (as reviewed in Section 2.3). This was done using the DampingFactor option in Mathematica. After trying different values between 1 and 0.001, there was no change in the converged solution. While Newton's method was able to converge to the correct solution, it succeeded or failed depending on the number of steps to take. For this system of constraints, it failed for one or two steps and it succeeded for three or more steps. These numbers would, of course, be different for any other system of constraints, requiring user intervention to select a proper number of iterations. Setting the number of iterations to an arbitrarily large number would, indeed, probably work for a large number of design cases, but it would be prohibitively time-consuming and one would never know if it failed without monitoring the convergence.

6. CONCLUSIONS

Variational design is a powerful paradigm for design, but has been limited in its application by the inability to characterize multiple solutions resulting from systems of nonlinear equations. Previous methods have required the user to either 1) manually guide the constraint solver routines, 2) manually select the correct solution from a list, or 3) input enough extra information so as to uniquely-constrain the solution. As opposed to these methods, we have introduced the notion of a solution trajectory and demonstrated algorithms for tracking the trajectory of a desired solution as the values of any external parameters change. An algorithm was presented and demonstrated to automatically perform reconvergence to a consistent solution.

We are excited about the future of these techniques in the area of variational design and continue to enhance the methods. Future work includes solving two special cases: First, if two or more solution trajectories intersect, the presence of multiple solutions will be detected and the algorithm cannot continue. Second, if a solution trajectory crosses a discontinuity (continuous in the complex plane, but discontinuous in the real plane), the algorithm cannot continue.   Also, we plan on extending the design consistency techniques to optimization problems (for functionality similar to DesignShell (Agrawal, 1991)). Finally, improvements in speed and accuracy will help make the algorithm more useful. We plan to pursue and extend interval analysis techniques for all these problems and we look forward to the kinds of applications that will result when design consistency can be assured in variational design.

7. ACKNOWLEDGMENTS

A great deal of respect and gratitude goes out to Leonid Charny whose technical leadership motivated and guided much of this research.

8. REFERENCES

R. Agrawal, 1991, A Constraint Management Approach for Optimal Design of Mechanical Systems, Ph.D. Thesis, Ohio State University.

R. Agrawal, G.L. Kinzel, R. Srinivasan, and K. Ishii, 1993, "Engineering Constraint Management Based on an Occurrence Matrix Approach", Journal of Mechanical Design, vol. 115, March, pp 103-109.

P.L. Beaty, P.A. Fitzhorn, and G.J. Herron, 1994, "Extensions in Variational Geometry that Generate and Modify Object Edges Composed of Rational Bezier Curves", Computer-Aided Design, vol. 26, no. 2, February, pp 98-108.

A. Borning, B. Freeman-Benson, M. Wilson, 1992, "Constraint Hierarchies", Lisp and Symbolic Computation, pp 223-270.

S. Alasdair Buchanan and Alan de Pennington, 1993, "Constraint Definition System: A Computer Algebra Based Approach to Solving Geometric Problems", Computer Aided Design, December, vol. 25, no. 12, pp. 741-750.

D.T. Gallaher, 1984, Variational Systems in Computer-Aided Design, M.S. Thesis, Massachusetts Institute of Technology.

R. Hammer, M. Hocks, U. Kulisch, and D. Ratz, 1993, Numerical Toolbox for Verified Computing I, Springer-Verlag, Germany.

R. Krawczyk and A. Neumaier, 1985, "Interval Slopes for Rational Functions and Associated Centered Forms", SIAM Journal of Numerical Analysis, vol. 22, no. 3, pp. 604-616.

R.A. Light, 1980, Symbolic Dimensioning in Computer-Aided Design, M.S. Thesis, Massachusetts Institute of Technology.

R. Light and D. Gossard, 1982, "Modification of Geometric Models through Variational Geometry", Computer-Aided Design, vol. 14, no. 4, July, pp 209-214.

V.C. Lin, 1981, Three-Dimensional Variational Geometry in Computer-Aided Design, M.S. Thesis, Massachusetts Institute of Technology.

J. Mackrell, 1993, "Making Sense of a Revolution", Computer Graphics World, November, pp 26-38.

O. Maimon and D. Braha, 1996, "On the Complexity of the Design Synthesis Problem", IEEE Journal of Systems, Man, and Cybernetics, January.

O. Maimon and D. Braha, 1995, "An Exploration of the Design Process", Technical Report, Boston University Dept. Of Manufacturing Engineering.

R.E. Moore, 1977, "A Test for Existence of Solutions to Nonlinear Systems", SIAM Journal of Numerical Analysis, vol. 14, no. 4, pp 611-615.

G. Nelson, 1985, "JUNO, A Constraint-Based Graphics System", ACM Transactions, vol. 19, no. 3, pp 235-243

A. Neumaier, 1989, "Rigorous Sensitivity Analysis for Parameter-Dependent Systems of Equations", Journal of Mathematical Analysis and Applications, vol. 144, pp 16-25.

A. Neumaier, 1990, Interval Methods for Systems of Equations, Cambridge University Press, New York, NY.

C. Padmanabhan and G.L. Kinzel, 1986, "Comparison of Different Non-Linear Equation Solvers for Mechanical Design Problems", Computers In Engineering, vol. 2, pp 233-237

S.M. Rump, 1990, "Rigorous Sensitivity Analysis for Systems of Linear and Nonlinear Equations", Mathematics of Computation, vol. 54, no. 190, April, pp 721-736.

D. Serrano, 1984, MathPAK: An Interactive Preliminary Design Package, M.S. Thesis, Massachusetts Institute of Technology.

N. Sridhar, R. Agrawal, and G.L. Kinzel, 1993, "Active Occurrence-Matrix-Based Approach to Design Decomposition", Computer-Aided Design, vol. 25, no. 8, August, pp 500-512.

M.A. Wilson, 1993, Hierarchical Constraint Logic Programming, Ph.D. Thesis, University of Washington, May.

Y. Yamaguchi and F. Kimura, 1987, "Interaction Management in CAD Systems with History Mechanism", Eurographics '87, Elsevier Science Publishers, pp 543-554.

APPENDIX A: INTERVAL ANALYSIS TECHNIQUES

This appendix gives a quick overview of interval analysis techniques.   More thorough coverage of interval analysis techniques is available in (Neumaier, 1990 and Hammer et al., 1993). The field of interval analysis has come up with many powerful methods for analyzing systems of equations when they are subject to change. These methods are used in MECHAD for maintaining a consistent solution trajectory when parameters of the systems of equations are changed.

- Iteratively improving an enclosure of a solution to a system of equations
- Rigorous sensitivity analysis of a system of equations
- Existence and uniqueness tests of systems of equations over intervals of parameters

Let [x] be an interval and let $\tilde{x}$ be any real number in interval [x],

$$[x] \equiv \left[\downarrow[x], \uparrow[x] \right] := \left\{ \tilde{x} \in \Re \mid \downarrow[x] \leq \tilde{x} \leq \uparrow[x] \right\}$$

Every interval, [x] , has a midpoint,

$$\hat{x} = \mathrm{mid}([x]) = (\uparrow[x] + \downarrow[x])/2$$

and a radius,

$$\mathrm{rad}([x]) = (\uparrow[x] - \downarrow[x])/2$$

The interior of a interval is denoted by

$$\mathrm{int}([x]) = \{\tilde{x} \in \Re \mid \downarrow[x] < \tilde{x} < \uparrow[x]\}$$

The hull of a nonempty bounded subset of $\Re$, S, is the tightest interval that encloses S and is denoted by
$Hull\{S\} = [\inf(S), \sup(S)]$.

Arithmetic operations are defined on intervals [x] and [y] as follows:

- $[x] \circ [y] =$

  $Hull\{\downarrow[x] \circ \downarrow[y], \downarrow[x] \circ \uparrow[y], \uparrow[x] \circ \downarrow[y], \uparrow[x] \circ \uparrow[y]\}$ for

  $\circ \in \{+, -, *, \backslash\}$

- $\varphi([x]) = [\varphi(\downarrow[x]), \varphi(\uparrow[x])]$ for $\varphi \in \{\mathrm{sqrt}, \exp, \ln\}$

- $\mathrm{sqr}([x]) = [0, [x]^2]$ if $0 \in [x]$ or $[\downarrow[x]^2, \uparrow[x]^2]$ otherwise

Other interval operations can be defined similarly.

A.1 Solving Systems Of Interval Equations

We wish to find a solution to the system of equations,

$$\underline{f}(\underline{x}) = \underline{0} \tag{11}$$

within a given interval, $\underline{x} \in [\underline{z}], [\underline{z}] \in I\Re^n$. Solution strategies can be adapted from the mean value theorem:

$$\underline{f}(\mathrm{mid}([\underline{z}])) - \underline{f}(\underline{x}^*) \in \underline{f}'([\underline{z}]) \cdot (\mathrm{mid}([\underline{z}]) - \underline{x}^*) \tag{12}$$

$$\text{for all } \underline{x}^* \in [\underline{z}]$$

Since $\underline{x}^*$ is a zero of $\underline{f}(\underline{x}) = \underline{0}$,

$$\underline{f}(\mathrm{mid}([\underline{z}])) \in \underline{f}'([\underline{z}]) \cdot (\mathrm{mid}([\underline{z}]) - \underline{x}^*) \tag{13}$$

It is commonly known that iteration methods such as this one work better if preconditioned by a real matrix (typically, the inverse of the midpoint matrix of $\underline{f}'([\underline{z}])$).

$$R \cdot \underline{f}(\mathrm{mid}([\underline{z}])) \in R \cdot \underline{f}'([\underline{z}]) \cdot (\mathrm{mid}([\underline{z}]) - \underline{x}^*) \tag{14}$$

If we let $A = R \cdot \underline{f}'(([\underline{z}]))$ and $\underline{b} = R \cdot \underline{f}(\mathrm{mid}([\underline{z}]))$ then (B.4) can be rewritten as

$$\underline{b} \in A \cdot (\mathrm{mid}([\underline{z}]) - \underline{x}^*) \tag{15}$$

and iteratively solved using Gauss Seidel iteration. Let $\underline{c} \approx \mathrm{mid}([\underline{z}]) \in [\underline{z}]$ and $\tilde{\underline{x}} \in [\underline{z}]$. The system is rewritten as $A \cdot (\underline{c} - \tilde{\underline{x}}) = \underline{b}$. Writing this system in terms of its individual components yields

$$\sum_{j=1}^{n} A_{ij} \cdot (\underline{c}_j - \tilde{\underline{x}}_j) = \underline{b}_i \tag{16}$$

and solving (16) for $\tilde{\underline{x}}_i$ gives:

$$\tilde{\underline{x}}_i \in \underline{c}_i - \left(\underline{b}_i + \sum_{\substack{j=1 \\ j \neq i}}^{n} A_{ij} \cdot (\underline{x}_j - \underline{c}_j)\right) / A_{ii} = [\underline{z}] \tag{17}$$

The Gauss-Seidel operator improves on (17), by observing that each iteration can use the previous interval, $\underline{y} = [\underline{z}]$, in place of x_j :

$$\underline{y} = [\underline{z}] \tag{18}$$

$$\underline{y}_i = \left(\underline{c}_i - \left(\underline{b}_i + \sum_{\substack{j=1 \\ j \neq i}}^{n} A_{ij} \cdot (y_j - \underline{c}_j)\right)\right) / A_{ii} \cap \underline{y}_i$$

Equation (18) is the Gauss-Seidel operator $\left(y = N_{GS}([\underline{z}])\right)$.

A.2 Existence And Uniqueness Of Solutions

Interval analysis techniques allow not only for the solution of systems of equations, but also theorems to check for the existence and the uniqueness of solutions within a given interval. The following theorem is proved by Neumaier (1990):

Let $F: D_0 \subseteq \Re^n \to \Re^n$ be countinuously differentiable on $D \subseteq D_0$. If $\tilde{\underline{x}} \in [\underline{z}] \in D$ and $\underline{x}'$ denotes the Gauss-Seidel operator, then:

   (i) Every zero $\underline{x}^* \in [\underline{z}]$ of F satisfies $\underline{x}^* \in \underline{x}'$.

   (ii) If $\underline{x}' \cap [\underline{z}] = 0$ then F contains no zero in x.

   (iii) If $\tilde{\underline{x}} \in \mathrm{int}([\underline{z}])$ and $0 \neq \underline{x}' \subseteq \mathrm{int}([\underline{z}])$ then F contains a unique zero in x.

APPENDIX B: CONSTRAINT MODEL OF BEAM

For the design of the beam, the following constraint model was utilized. Every attribute is ultimately expressed as a function of the dimensions.

B.1 Dimensions

L,	Length of beam (m)
w,	Width of beam top (m)
t_1,	Height of beam top (m)
t_2,	Height of beam bottom (m)
t_3,	Width of beam bottom (m)
σ_{max},	Ultimate stress (Pa)
n,	Safety factor
E,	Young's Modulus (Pa)
ρ,	Material density ($kg \cdot m^3$)

B.2 Attributes

Volume:

$$L \cdot \left(t_2 t_3 + t_1 w\right)$$

Mass:

$$\text{Volume * Density} = \rho \cdot L \cdot \left(t_2 t_3 + t_1 w\right)$$

Total Surface Area:

$$2L\left(t_1 + t_2 + w\right) + t_2 t_3 + t_1 w$$

Maximum Moment:

$$M_{max} = \frac{\sigma_{max} I}{c} = \frac{\sigma_{max} t_2^{\,3} t_3 \left(t_2 t_3 + t_1 w\right)}{12 \cdot \left(\frac{t_2^{\,2} t_3}{2} + w t_1 \left(\frac{t_1}{2} + t_2\right)\right)}$$

Maximum Load:

$$P_{max} = \frac{2 M_{max}}{L^2} = \frac{\sigma_{max} t_2^{\,3} t_3 \left(t_2 t_3 + t_1 w\right)}{3L^2 \left(t_2^{\,2} t_3 + t_1^{\,2} w + 2 t_1 t_2 w\right)}$$

Maximum Allowable Load:

$$P_{allow} = \frac{2 M_{max}}{n L^2} = \frac{\sigma_{max} t_2^{\,3} t_3 \left(t_2 t_3 + t_1 w\right)}{3 n L^2 \left(t_2^{\,2} t_3 + t_1^{\,2} w + 2 t_1 t_2 w\right)}$$

End Angle at Maximum Load:

$$\theta_{max} = \frac{P_{max} L^3}{6EI} =$$

$$\frac{2 L \sigma_{max} t_2^{\,3} t_3 \left(t_2 t_3 + t_1 w\right)^2}{\left(\begin{array}{c} 3E\left(t_2^{\,2} t_3 + t_1^{\,2} w + 2 t_1 t_2 w\right) \cdot \\ \left(t_2^{\,4} t_3^{\,2} + 4 t_1^{\,3} t_2 t_3 w + 6 t_1^{\,2} t_2^{\,2} t_3 w + 4 t_1 t_2^{\,3} t_3 w + t_1^{\,4} w^2\right) \end{array}\right)}$$

End Angle at Maximum Allowable Load:

$$\theta_{allow} = \frac{P_{all} L^3}{6EI} =$$

$$\frac{2 L \sigma_{max} t_2^{\,3} t_3 \left(t_2 t_3 + t_1 w\right)^2}{\left(\begin{array}{c} 3nE\left(t_2^{\,2} t_3 + t_1^{\,2} w + 2 t_1 t_2 w\right) \cdot \\ \left(t_2^{\,4} t_3^{\,2} + 4 t_1^{\,3} t_2 t_3 w + 6 t_1^{\,2} t_2^{\,2} t_3 w + 4 t_1 t_2^{\,3} t_3 w + t_1^{\,4} w^2\right) \end{array}\right)}$$

End Deflection at Maximum Load:

$$\Delta_{max} = \frac{P_{max} L^4}{8EI} =$$

$$\frac{L^2 \sigma_{max} t_2^{\,3} t_3 \left(t_2 t_3 + t_1 w\right)^2}{\left(\begin{array}{c} 2E\left(t_2^{\,2} t_3 + t_1^{\,2} w + 2 t_1 t_2 w\right) \cdot \\ \left(t_2^{\,4} t_3^{\,2} + 4 t_1^{\,3} t_2 t_3 w + 6 t_1^{\,2} t_2^{\,2} t_3 w + 4 t_1 t_2^{\,3} t_3 w + t_1^{\,4} w^2\right) \end{array}\right)}$$

End Deflection at Maximum Allowable Load:

$$\Delta_{allow} = \frac{P_{all} L^4}{8EI} =$$

$$\frac{L^2 \sigma_{max} t_2^{\,3} t_3 \left(t_2 t_3 + t_1 w\right)^2}{\left(\begin{array}{c} 2nE\left(t_2^{\,2} t_3 + t_1^{\,2} w + 2 t_1 t_2 w\right) \cdot \\ \left(t_2^{\,4} t_3^{\,2} + 4 t_1^{\,3} t_2 t_3 w + 6 t_1^{\,2} t_2^{\,2} t_3 w + 4 t_1 t_2^{\,3} t_3 w + t_1^{\,4} w^2\right) \end{array}\right)}$$

APPENDIX C: FORMAL DESIGN PARADIGM DESCRIPTION

Having developed the method, we are ready now to formally present the full design paradigm. As there are many different approaches to design, it is important to formally define the specific design paradigm that MECHAD assumes. Maimon and Braha (1995) have derived a formal description of an evolutionary design model. In this section, we use their model to formally define the design paradigm behind the implementation of MECHAD.

Maimon and Braha describe a design *execution* as a sequence of synthesis states, each state being characterized by a design pair, $\langle \theta, m \rangle$, where:

- θ is the current design specification and
- m is the current design realization.

Each transition between synthesis states is described by either termination (success or failure) or modification (of the specification or the design).

The design process can thus be represented as a sequence of synthesis states, $\{\langle \theta_0, m_0 \rangle, \langle \theta_1, m_1 \rangle, ..., \langle \theta_k, m_k \rangle\}$. Maimon and Braha denote the model of a general evolutionary design process as a tuple, $\langle DA, L, G_\theta, G_m \rangle$, where

- DA is the design representation scheme,
- L is the specification language,
- G_θ is a non-deterministic finite state machine that describes all the possible design specifications.
- G_m is a non-deterministic finite state machine that describes all the possible design artifacts, and

MECHAD can be shown to be an instance of the general evolutionary process model with the definitions given as follows.

C.1 DESIGN ARTIFACT CREATION, *DA*

The only design artifacts in MECHAD are those created as calculated from the specifications, θ. As such, the artifact space, DA, is defined by the tuple, $\langle M_0, C^0, M^* \rangle$ where M_0 is the set of atomic (or primitive) designs, $C^0 = $ the set of connectors from which composite objects can be defined, and from those, M^* is the set of all possible design artifacts. Currently in MECHAD, $M_0 = \mathfrak{R}^n$, $C^0 = \{\varnothing\}$, and $M^* = M_0 \in \mathfrak{R}^n$,

C.2 SPECIFICATION LANGUAGE, *L*

L is equal to the design specification language in MECHAD. At this point in the development, L, is not a full propositional specification language as recommended by Maimon and Braha. While the system of constraints can be considered to be a logical conjunction, disjunctions and negations are not currently supported.

Actual design execution takes place by selecting attributes of the design and defining their values. As MECHAD only solves fully-constrained systems, the user must enter the same number of constraints as there are dimensions. Under- or over- constrained systems are not allowed.

C.3 SPECIFICATION FSM, G_θ

G_θ is defined by the tuple, $\langle L, L^0, M^*, T_{G_\theta} \rangle$ where:

- L denotes the set of all possible states of G_θ. In MECHAD, L is defined by the tuple, $\langle \underline{f}, \underline{b}, \underline{Q}, \underline{q} \rangle$ where $\underline{f}$ is the set of attributes or dimensions that the user is constraining, $\underline{b}$ is the vector of corresponding user-defined values $\left(\underline{f}(\underline{d}) = \underline{b} \right)$, $\underline{Q}$ is the set of mathematical expressions defining the qualitative design specifications, and $\underline{q}$ is the vector of corresponding bounds $\left(\underline{Q}(\underline{d}) \geq \underline{q} \right)$. Although the user defined values are real numbers ($L \in \Re^n$), finiteness of the domain is a reasonable assumption here since the values of the dimensions can be quantized at the level of the desired tolerance;

- L^0 is the set of qualitative specifications and the initial set of constraints entered by the user $\left(\langle \underline{f}^0, \underline{b}^0, \underline{Q}, \underline{q} \rangle \right)$;

- M^* is the set of all possible design specifications (as given by G_m); and

- T_{G_θ} is the state transition mapping. The transitions between states are triggered by any user modification of the state tuple. T_{G_θ} terminates if the specification satisfies the design (for every $\theta_k \in L$, if θ_k satisfies m, then $T_{G_\theta}(\theta_k, m) = \varnothing$).

C.4 DESIGN FSM, G_m

G_m is defined by the tuple, $\langle M^*, M^0, L, T_{G_m} \rangle$ where:

- M^* denotes the states of the design FSM. It is the same as defined for DA. Finiteness of the domain due to tolerance limitations is assumed;

- M^0 is the initial state of the design process, $\left(M^0 = \left\{ \underline{d}^0 \mid \underline{f}^0 \left(\underline{d}^0 \right) = \underline{b}^0 \right\} \right)$. User intervention may be needed to help define the initial state (in the presence of multiple feasible solutions);

- L is the set of possible specification parts (as defined for G_θ); and

- T_{G_m} is the state transition mapping. Transitions between states are only triggered by a change in specification resulting in a new solution. T_{G_m} terminates if the specification satisfies the design (for every $m_k \in M^*$, if m_k satisfies θ, then $T_{G_m}(m_k, \theta) = \varnothing$).

**DE-Vol. 87, Reliability, Stress Analysis, and Failure Prevention
Issues in Emerging Technologies and Materials
ASME 1995**

THE CONCEPTUAL DESIGN OF
ROTO-COOLER AIR CONDITIONING SYSTEM

Yuan Mao Huang

Department of Mechanical Engineering
National Taiwan University
Taiwan
Republic Of China

ABSTRACT

The conventional air conditioning systems have many shortcomings. This paper presents the conceptual design of roto-cooler air conditioning system. With the specification of air conditioning systems, the design goal is abstracted into a phrase. The overall function and the sub-function structure of the system are constructed. The physical effects, the physical principles and the solution principles of the sub-functions and the combined solution principles are generated. The weighted values and assigned grade points of the standard items and the weighted grades of the proposed combined solution principles of system are calculated. Therefore, a roto-cooler air conditioning system can be determined.

INTRODUCTION

The conventional air conditioning systems are complex, large, heavy, expensive, not easy to maintain, and their coefficients of performance per weight are low. The high pressure line is required to supply the high pressure refrigerant. Due to the vibration of the system and the roughness of the road condition, it may cause leakage of the refrigerant or entrain the dirt particles into the system. The temperature of the evaporator can not drop below 0 C. Otherwise, water vapor will condense on the surface of evaporator and the performance of system will be degraded. The cooling down time is relatively long. Especially, the pollution due to the refrigerant may cause a serious effect on the atmospheric condition. The harmful ultra-violet rays of the sun can reach the earth directly to result in increasing the human skin problems and rising the atmospheric temperature. The roto-cooler air conditioning systems may challenge the conventional air conditioning systems (Edwards and McDonald, 1972), (Edwards, 1975). The purpose of this project is to study the conceptual design of roto-cooler air conditioning systems (Pahl &Beitz,1988), (Cross, 1994), (Ullman, 1992) and to understand the advantages of developing the roto-cooler air conditioning system.

METHOD OF APPROACH
Specification

The specification of air conditioning systems is proposed by various potential buyers and manufactures as shown in Table 1. For the conceptual design of systems, requirements and constraints are divided into three catagories; M, D and W, which stand for the requirements and constraints that must, demand or wish, respectively, to be satisfied.

Abstraction

After understanding the specification, the design goal can be abstracted and condensed into a phrase as "to design an air conditioning system with the high cooling rate, fast cooling, high coefficient of performance but less expensive, small size and the minimum deterioration to the atmospheric condition".

Function Structure

Based on the concept of system, the input and output of the energy, mass and signal crossing the boundary of the system, the overall function structure of an air conditioning system is shown in Figure 1.

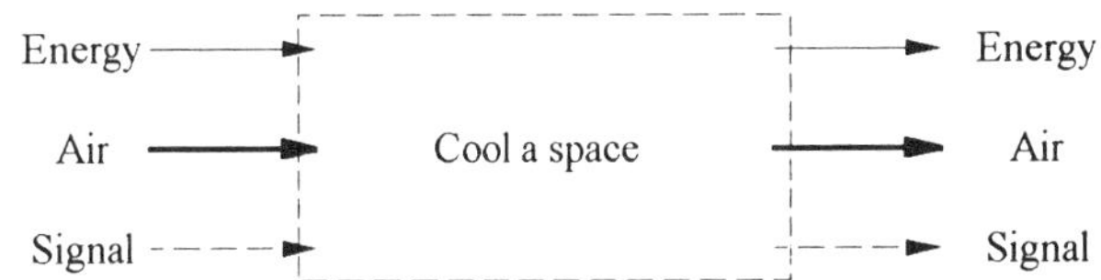

FIGURE 1 THE OVERALL FUNCTION STRUCTURE OF
AIR CONDITIONING SYSTEM

TABLE 1   THE SPECIFICATION OF AIR CONDITIONING SYSTEMS

Date of Change	D,M, or W	Specification for air conditioning systems	Page
		Requirements and Constraints	**Responsible**
		energy	
	M	less energy	
	M	high coefficient of performance	
		signal	
	D	automatic control temperature	
	D	automatic control cooling time	
		geometry	
	D	small size	
	W	light weight	
		performance	
	D	cooling room larger than 6 m x 6 m x 2.44 m	
	M	cool four persons inside the room and dissipated heat rate of every person is 63w	
	M	room temperature remains 25.6 C	
	M	cooling capacity 1.52 kw	
	D	cooling down time less than 3.5 minutes	
		life	
	M	endurance and long life	
		safety	
	M	no harm to users during operation	
	M	no harm to system during operation	
		assembly	
	W	easy to be mounted	
		operation	
	D	quiet	
	M	operation for 24 hours/day	
		maintenance	
	W	long period of maintenance	
	W	easy to maintain	
		expenditure	
	D	less expensive	
		environment	
	M	satisfactory of environmental protection	
		Project Manager:	Issued Date:

Sub-function Structure

The function structure can be remodelled and reconstructed with the sub-functions as shown in Figure 2. The sub-functions in the blocks surrounded by the solid lines and the dashed lines are the main sub-functions and the auxiliary sub-functions, respectively.

Solution Principles

There are many solution principles to satisfy each sub-function. Based on the physical effects and the physical principles, some solution principles for each sub-function are listed in Table 2. The solution principles of all sub-functions are combined. Some illegible combined solution principles are eliminated. The first combined solution principles, solution A, to satisfy the overall function is 1-1, 2-1, 2-2, 3-1 and 3-8. The vapor compressive air conditioning system is such a system (Reynolds, 1978). The combined solution B to satisfy the overall function is 1-1, 2-1, 2-2, 3-1 and 3-8. The vacuum refrigeration system is such a system (Reynolds, 1978). The combined solution C to satisfy the overall function is 1-1, 2-1, 2-3, 3-1 and 3-8. The absorption refrigeration cycle system is such a system (Reynolds, 1978). The combined solution D to satisfy the overall function is 1-6, 2-1 and 3-8. The roto-cooler air conditioning system with an elliptical stator is such a system.

EVALUATION

Some standard items are required to evaluate the feasibility of the systems. The standard items chosen for evaluation are listed in Table 3. The effect of all items can be divided into the

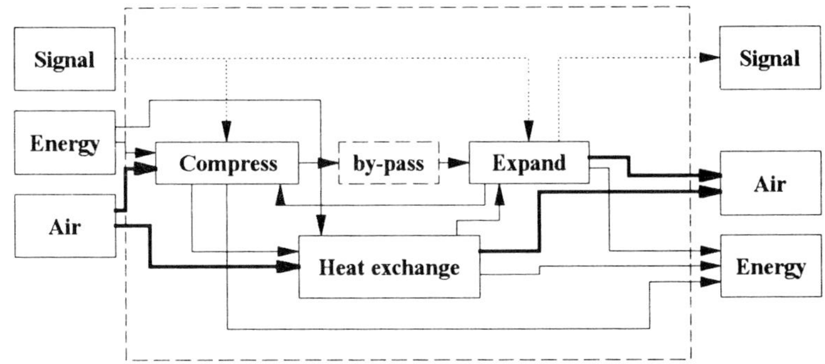

FIGURE 2 THE SUB- FUNCTION STRUCTURE OF AIR CONDITIONING SYSTEM

TABLE 2 THE SUB-FUNCTIONS, PHYSICAL EFFECTS, PHYSICAL PRINCIPLES AND SOLUTION PRINCIPLES

sub-function	physical effect	physical principle	solution principle
1. compress	change volume can change pressure	$pv = nRT$	1-1 reciprocating compressor 1-2 reservoir compressor 1-3 rotary compressor 1-4 eccentric rotary compressor 1-5 multiple-disk centrifugal compressor 1-6 elliptical stator rotary compressor 1-7 axial flow compressor
2. exchange heat	heat transfer from high temperature to low temperature	$\dot{Q} = UA\,\Delta T$ or $\dot{Q} = f(\Delta T)$	2-1 enlength duct, install fin increase duct diameter 2-2 heat conduction 2-3 heat convection 2-4 heat radiation
3. expand	change volume can change pressure	$pv = nRT$	3-1 to 3-7 reversed processes of 1-1to 1-7 3-8 expansion value

economic consideration and the technical consideration. The object tree is built for comparison of the relative importance of all standard items as shown in Figure 3. Based on the statistical result from a survey, the weighted values are determined as shown in Figure 4. Using the VDI 2255 method (Pahl & Beitz, 1988) and grading system from 0 to 4 points for all standard items in evaluation, the best performance is assigned to have 4 points and the worst one is assigned to have 0 point for V_i as shown in Table 4. The weighted value C is multiplied by the graded point of the the standard item V_i to obtain the weighted grade W_i. The summation of weighted grade W_i is used for comparison to determine the best feasibility of the systems.

The schematic drawing of the rotor-cooler air conditioning system shown in Figure 5 is comparable to other current existing systems. The stator is an elliptical shape. The rotor consists of ten slots and the blades can slide in the slots.

For the counterclockwise rotation of the rotor, air at essentially atmospheric pressure, P_a, and temperature, T_a, is drawn from the inlet duct into the blade segment V_1 as shown in Figure 5. Air

properties inside the blade segment are compressed. Most of air is then pumped from the blade segment into the heat exchanger where air is cooled down by rejecting heat to surrounding air. Similarly, air is assumed to be flowing into the expansion blade segment from the outlet of the heat exchanger. Air is then pumped through the outlet duct into the space being cooled. This is the reversed Brayton cycle and the air cycle refrigeration. The elliptical stator encloses the whole system as a unit. Therefore, the sytem is simple and easy to maintain. The coefficient of performance per unit weight is higher than those of the conventional air conditioning systems. No refrigerant is needed. The cooling down time is relatively short.

Similarly, the specification of the roto-cooler air conditioning system is shown in Table 5. The sub-function structure of the system is shown in Figure 6. The solution principles of the sub-functions are shown in Table 6.

The solution A can be concluded as follows:

1. change the air inlet angle to have the maximum blade segment volume
2. change the air outlet angle to have the maximum blade

symbol	standard items
a	fast cooling
b	high coefficient of performance
c	less expensive
d	small size and light weight
e	easy to maintain
f	endurance and long life
g	safe
h	long period between maintenance
i	quiet
j	easy to mount
k	multi-functions
l	automatic control

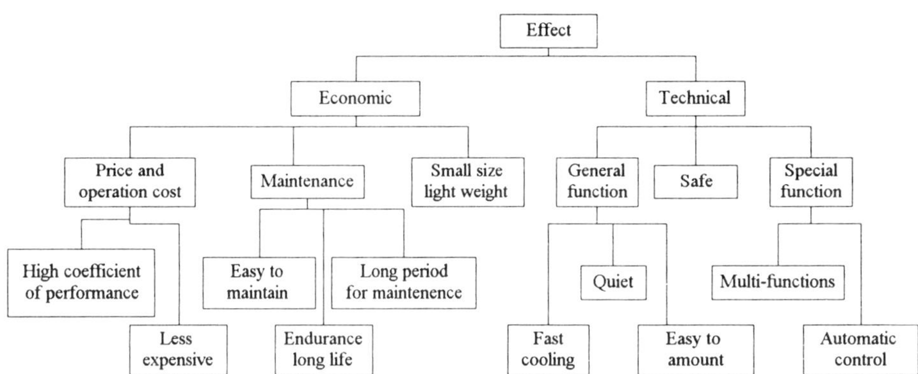

FIGURE 3 THE OBJECT TREE FOR EVALUTION

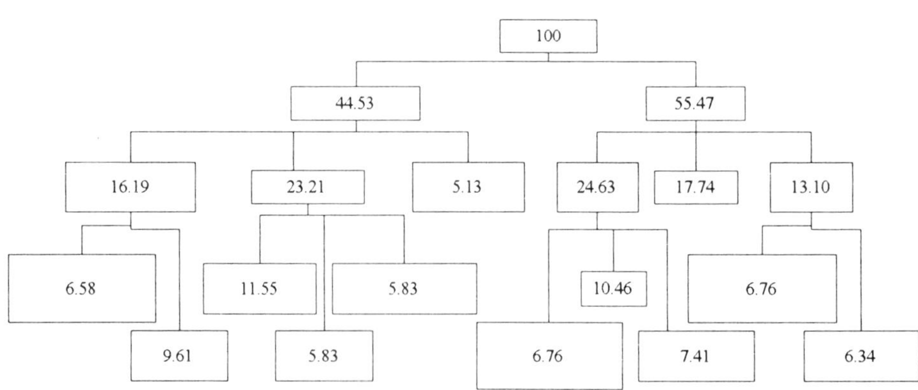

FIGURE 4 THE WEIGHTED VALUES OF ALL ITEMS FOR EVALUTION

TABLE 4 COMPARISON OF SOLUTIONS

standard items		solution A		solution B		solution C		solution D	
symbol	Ci	Vi	Wi	Vi	Wi	Vi	Wi	Vi	Wi
a	6.76	4	27.04	3	20.28	3	20.28	4	27.04
b	6.58	3	19.74	4	26.32	3	19.74	1	6.58
c	9.61	4	38.44	0	0	2	19.22	3	28.83
d	5.13	4	20.52	1	5.13	2	10.26	3	15.39
e	11.55	3	34.65	1	11.55	2	23.10	3	34.65
f	5.83	3	17.49	3	17.49	3	17.49	3	17.49
g	17.74	2	35.48	2	35.48	2	35.48	3	53.22
h	5.83	2	11.66	2	11.66	2	11.66	3	17.49
i	10.46	3	31.38	3	10.46	3	10.46	2	20.93
j	7.41	3	22.23	3	22.23	3	22.23	3	22.23
k	6.76	2	13.52	2	13.52	2	13.52	1	6.76
l	6.34	1	6.34	1	6.34	1	6.34	1	6.34
sum	100	28	278.4	25	180.46	28	209.7	30	256.9

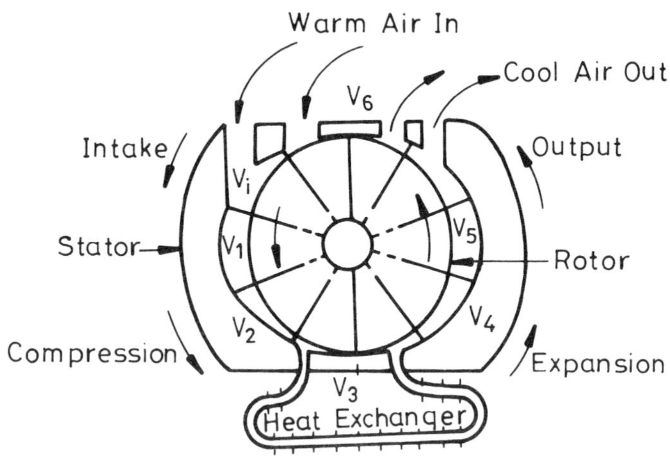

FIGURE 5 THE SCHEMATIC DRAWING OF THE ROTO-COOLER AIR CONDITIONING SYSTEM

segment volume

3. use the rotor with the radius same as the semi-minor axis length of ellipse
4. change the location of the heat exchanger input and output port locations
5. use an ellipse with different eccentricities
6. install fins
7. use self-lubricated material
8. use the different shapes of blades and number of blades
9. reduce the force acted on the bottom of the blade
10. keep certain clearances between the rotor and the stator and the covers on both sides
11. reduce the roughness of the contact surface with air

The other solutions are as followed:

Solution B: same as the solution A, except the item 2 is replaced by 5-1 in Table 6.

Solution C: same as the solution A, except the item 7 is replaced by 6-1-2 in Table 6.

Solution D: same as the solution A, except the item 7 is replaced by 6-1-3 in Table 6.

Solution E: same as the solution A, except the item 10 is replaced by 6-3-1 in Table 6.

Solution F: same as the solution A, except the item 10 is replaced by 6-3-2 in Table 6.

Three types of blades such as the radial blade, the inclined blade and the circular arc blade are proposed as shown in Figure 7. The roto-cooler air conditioning system with the radial blades is shown in Figure 5. The compressors with the eccentric rotror and the radial blades and the inclined blades are shown in the compressor handbook (1979). Using the standard items in Table 3 and the weighted values for all standard items in Figure 4, the weighted grades are shown in Table 7.

TABLE 5 THE SPECIFICATION OF ROTO-COOLER AIR CONDITIONING SYSTEM

| | | Specification | | Page |
| | | for   roto-cooler air conditioning system | | |
Date of Change	D,M, or W	Requirements and Constraints		Responsible
		energy		
	M	less energy		
	M	high coefficient of performance		
		signal		
	D	automatic control temperature		
	D	automatic control cooling time		
		geometry		
	D	size of stator; major axis length <15.24 cm;		
	W	minor axis length <11.68 cm;		
	W	depth to be 15.2476 cm		
	D	heat exchanger length longer 7.62 m		
	D	small size		
	W	light weight		
		performance		
	W	coefficient of performance higher than 0.4		
	W	thermal coefficient of performance higher than 1.13		
	D	cooling room larger than 6 m x 6 m x 2.44 m		
	M	four persons inside the room and dissipated heat rate of every person is 63w		
	M	room temperature remains 25.6 C		
	M	minimum cooling capacity 1.52 kw		
	D	cooling down time less than 3.5 minutes		
		life		
	M	endurance and long life		
		safety		
	M	no harm to users during operation		
	M	no harm to system during operation		
		assembly		
	W	easy to be mounted		
		operation		
	D	quiet		
	M	operation for 24 hours/day		
		maintenance		
	W	long period of maintenance		
	W	easy to maintain		
		expenditure		
	D	less expensive		
		environment		
		satisfactory environmental protection		
		Project Manager:		Issued Date:

DISCUSSION

The weighted values of all standard items as shown in Figure 4 are obtained based on a small scale survey of 100 persons for general products, which are not specified in this preliminary study to save time and cost. The roto-cooler air conditioning system shows its potential from the result shown in Table 4. The system with the radial blades should be developed based on the result shown in Table 7.

CONCLUSION

The roto-cooler air conditioning system has the potential for further development based on this preliminary study of the conceptual design of air conditioning systems. Although the weighted values and the grades for the standard items which are not based on the statistical data of the roto-cooler air conditioning system are determined, this study provides an example for the conceptual design of air conditioning systems. As long as the

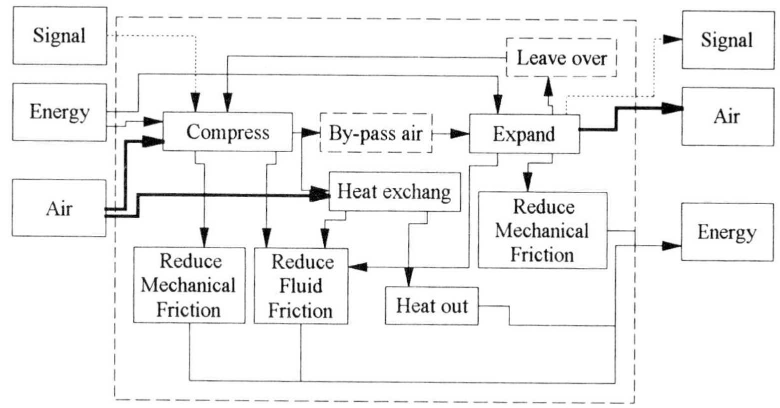

FIGURE 6 THE SUB-FUNCTIONS OF ROTO-COOLER AIR CONDITIONING SYSTEM

weighted values and the grades for the standard items of the specific systems are reasonable and dependable, a better system can be determined.

ACKNOWLEDGMENT

The author would like to express his sincere thank to the National Science Council of the Republic of China for the Grant number 82-0401-E-002-425 to complete this study.

REFERENCE

Carrier Air Conditioning Company, 1986, Handbook of Air Conditioning System Design, McGraw Hill.

Cross, N. , 1994, ENGINEERING DESIGN METHOD - Strategies for Product Design, John Wiley & Sons.

Edwards, T. C. , 1975, "The ROVAC Automotive Air Conditioning System" SAE paper 750403.

Edwards, T. C. And McDonald, A. T., 1972, "ROVACS; A New Rotary Van Air-Cycle Air Conditioning And Refrigeration System" SAE paper 720079.

Gulf Publishing Company, 1979, Compressor Handbook for the Hydrocarbon Processing Industries, Gulf Publishing Company Book Division.

Pahl, G. & Beitz, W., 1988, Engineering Design - A System Approach, edited by Wallace, K., Springer-Verlag,.`

Reynolds, W. C., 1978, Engineering Thermodynamics, McGraw-Hill, pp 317-325, 330-335 and 344-3347.

Ullman, D. G., 1992, THE MECHANICAL DESIGN PROCESS, McGraw-Hill, Inc.

TABLE 6 THE SUB-FUNCTIONS, PHYSICAL EFFECTS, PHYSICAL PRINCIPLES AND
SOLUTION PRINCIPLES OF THE ROTO-COOLER AIR CONDITIONING SYSTEM

Sub-function	physical effect	Physical principle	Solution Principle
1. compress	change volume can change pressure	$pv = nRT$	1-1 change location of air input to have maximum blade segment volume 1-2 change location of heat exchanger inlet 1-3 use rotor with different radius 1-4 use ellipse with different eccentricity 1-5 use different type of blades
2. reduce bypass air	change volume can change quantity of air	$v = LA$	2-1 change heat exchanger inlet location 2-2 reduce air pressure at heat exchanger inlet 2-3 use rotor with radius same as semi-minor axis length
3. take heat out	heat transfers from high temperature to low temperature	$\dot{Q} = AU\Delta T$	3-1 increase pipe length 3-2 increase pipe diameter 3-3 install fin 3-4 use high thermal conductivity material 3-6 increase temperature difference between inner and outer heat exchanger surface
4. expand	change volume can change pressure	$pv = nRT$	4-1 change port location to increase blade segment volume 4-2 change output port location of heat exchanger 4-3 use rotor with different radii 4-4 use ellipse with different eccentricities 4-5 use different types of blades
5. reduce leavings of air	change volume can change quantity of air	$v = LA$	5-1 change output port location to have minimum blade segment volume 5-2 use rotor with radius same as semi-minor axis length of ellipse
6.reduce mechanical friction	reduce normal force can reduce friction	$F = \mu N$ 1. at stator	6-1-1 use self-lubricated material 6-1-2 use lubricant 6-1-3 use roller to replace the sliding friction by rolling friction 6-1-4 use different types of blades 6-1-5 reduce spring force acted on blade bottom
		2. at slot	6-2-1 to 6-2-4 same as 6-1-1 to 6-1-4 6-2-5 reduce difference of air pressures on both sides of blade
		3. at side cover	6-3-1 use self-lubricated material 6-3-2 use lubricant 6-3-3 use roller to replace sliding friction by rolling friction 6-3-4 to remain certain clearances between rotor and stator and both side covers
7. reduce fluid friction	particle contacts create friction	$\tau = \mu \dfrac{V}{y}$	7-1-1 reduce viscosity 7-1-2 reduce roughness of surface of material contacted with air

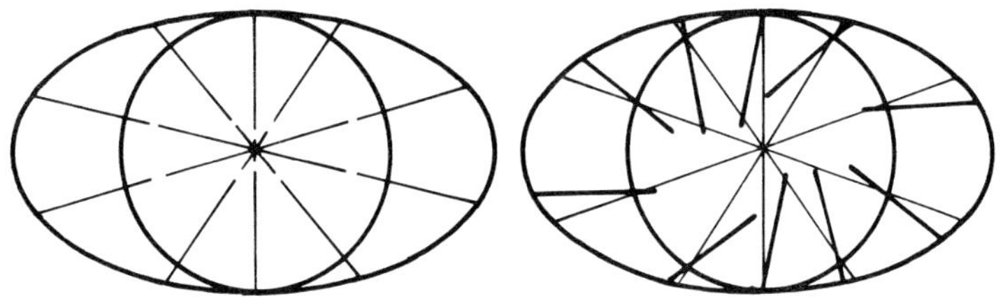

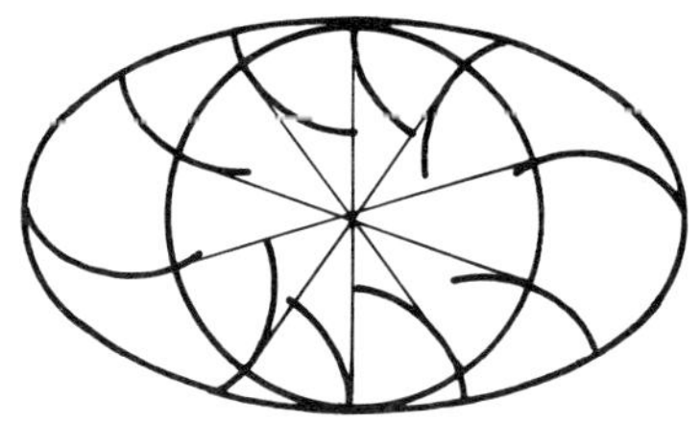

FIGURE 7 THREE TYPES OF BLANDS

TABLE 7 THE WEIGHTED GRADES FOR COMPARISON

standard items		solution											
		A		B		C		D		E		F	
symbol	Ci	Vi	Wi	Vi	Wi	Vi	Wi	Vi	Wi	Vi	Wi	Vi	Wi
a	6.76	4	27.0	0	0	4	27.0	4	27.0	4	27.0	4	27.0
b	6.58	4	26.3	0	0	4	26.3	4	26.3	4	26.3	4	26.3
c	9.61	4	38.4	0	0	4	38.4	4	38.4	2	18.4	4	38.4
d	5.13	4	20.4	4	20.4	3	15.4	3	15.4	4	20.4	4	20.4
e	11.55	4	46.2	3	34.7	3	34.7	3	34.7	3	34.7	3	34.7
f	5.83	3	17.5	3	17.5	1	5.8	4	23.3	1	5.83	2	11.7
g	17.74	3	53.2	3	53.2	1	17.7	4	71.0	1	17.7	2	35.5
h	5.83	0	0	4	23.3	0	0	0	0	0	0	0	0
i	10.46	3	31.4	3	31.4	1	10.5	4	41.8	1	10.5	2	20.9
j	7.41	4	29.6	3	22.2	3	22.2	3	22.2	4	29.6	4	29.6
k	6.76	4	27.0	0	0	2	13.5	2	13.5	2	13.5	2	13.5
l	6.34	4	25.4	4	25.4	4	25.4	4	25.4	4	25.4	4	25.4
sum	100	41	342.	27	228.	30	237	39	339.	30	229.	35	283.

FAILURE PATTERN OF MACHINING CENTERS (MC)

Jia Yazhou, Wang Molin
Mechanical Engineering Department
Jilin University of Technology
Changchun 130025
P. R. of China

Jia Zhixin
Mechanical Engineering Department
Harbin Institute of Technology
Harbin 150001
P. R. of China

ABSTRACT

This paper discusses the failure pattern and reliability of cutter-changeable CNC machine tools (machining centers) in China. 24 machining centers were tested and researched in the field over a period of two years. The failure data collected in the field are collated and statistically analysed. The results indicate that the failure distribution of MC fits the exponential distribution. This provides the basis for estimating the reliability of machining centers.

Key words: failure pattern, machining center, exponential distribution, field test

INTRODUCTION

The computerized numerical control (CNC) machine tool is the main equipment for modern metal machining. With the improvement of automatics, many machining centers (MC) which have cutter-base and manipulator and can change cutter automatically are emerging. However, the constitution of machine tools is getting more and more complex as their function extends. Consequently the failure probability increases in operation and the issue of reliability becomes important. But up till now, there has been little research of failure pattern and reliability of MC [1,2]. Consequently, frequent failures seriously hamper their broad use.

Since the failure distribution of MC has not been determined, a theoretical basis for reliability characterizations has not been obtained. Thus it is clear that failure analysis of MC is of critical importance.

The main purpose of this paper is to offer a further discussion about the statistical distribution of the MC's field failure and to set up a mathematical model of the failure distribution through theoretical deduction and verifying.

PRELIMINARY FITTING

A large amount of failure informations were collected and collated when the 24 MCs were tested and researched in the field over a period of two yesrs [3]. The distribution of failure time is fitted in order to discover the failure law.

Let $t_1, t_2, \cdots, t_n$ denote the observed value of MC's failure time array sequenced from small to large, that is $t_1 \leqslant t_2 \leqslant \cdots \leqslant t_n$. For any time $t > 0$, we define [4]

$$F(t_i) = \frac{i - 0.3}{n + 0.4} \qquad (i = 1, 2, \cdots, n) \qquad (1)$$

as expression for computing the median rank. Using this expression we can approximately estimate the cumulative distribution function of failure time of MC. In Equation (1) i denotes the original failure number and n the total number of failures. The distribution point diagram of MC's failure data is plotted in Figure 1 on the basis of the field records and Equation (1) along with the distribution curve.

Figure 1a shows the distribution point diagram of a certain MC, Figure 1b represents the statistical results for six other MCs. Failure analysis of 24 MCs has shown that failure distributions of the others reduce to similar results. In the figures, the abscissa indicates the time of MC's failure, t (hours); the ordinate indicates the distribution function $F(t)$. Relationship of the distribution function and the probability density function is given as follows:

$$f(t) = \frac{dF(t)}{dt} \quad . \tag{2}$$

From the fitting diagram, the shape of the failure distribution function $F(t)$ of MC appears convex and has no elbow, so we can conclude that its probability density function is not bell shaped, that is to say, it is not the normal distribution or the log normal distribution.

As it is known, the Weibull distribution is most useful for the lifetime test [5], with its cumulative distribution function given by

$$F(t) = 1 - exp[-(\frac{t}{\theta})^b] \tag{3}$$

where b is the shape parameter (Weibull gradient); θ is the characteristic life parameter. The Weibull distribution is actually a family of probability density functions. It may be noted that this function represents a simple exponential distribution when $b = 1$, a Rayleigh distribution when $b = 2$, and a good approximation to the Gaussian distribution when $b = 3.57$. At first, we fitted the field failure data with the Weibull distribution.

Rewrite Eg. (3) and take the logarithm, then

$$lnln\frac{1}{1 - F(t)} = blnt - bln\theta \quad . \tag{4}$$

Usually one can use median rank (see Eq. (1)) to evaluate $F(t)$ approximately. Then, we can construct the weibull probability distribution according to Eq. (4) by plotting the couples $(lnt_i, lnln\frac{1}{1-F(t_i)})$. We plot field failure data of MC on Weibull paper. Fig. 2a shows the result of plotting the failure data of the above mentioned individual machine tool, Fig. 2b shows the result for six machine tools. Here the lines are fitted by the linear regressive method. Using a special software package, the distribution coefficient, the linear regressive parameters and the correlation parameters for statistical testing are worked out. We introduce the notation [6]

$$S_{xx} = \sum_{i=1}^{n} (x_i - \bar{x})^2$$

$$S_{yy} = \sum_{i=1}^{n} (y_i - \bar{y})^2$$

$$S_{xy} = \sum_{i=1}^{n} (x_i - \bar{x})(y_i - \bar{y})$$

where

$$\bar{x} = \frac{1}{n} \sum_{i=1}^{n} x_i \quad , \quad \bar{y} = \frac{1}{n} \sum_{i=1}^{n} y_i \quad .$$

The linear regressive equation is given by

$$\hat{y} = \hat{a} + \hat{b}x \tag{5}$$

where $\hat{a}$ and $\hat{b}$ are estimators of the linear regressive parameters,

$$\hat{b} = \frac{S_{xy}}{S_{xx}} \tag{6}$$

and

$$\hat{a} = \bar{y} - \hat{b}\bar{x} \quad . \tag{7}$$

The goodness of fit is evaluated by correlation coefficient [6]

$$R = S_{xy} / [S_{xx} \cdot S_{yy}]^{\frac{1}{2}} \quad . \tag{8}$$

Note that the absolute value of correlation coefficient is $|R| < 1$, and the larger the value of $|R|$, the more significant the linear relation is. There is a critical value R_a according to each significance level α respectively, and it depends on the degree of freedom $(n - 2)$ only. Set $\alpha = 0.01$, the critical value of correlation coefficient can be obtained by

$$R_a = 2.573 / (\gamma + 3)^{\frac{1}{2}} \tag{9}$$

where $\gamma = n - 2$. In general, if $|R| > R_a$, the linear relation is significant.

Rewrite Eq. (4), then

$$y = bx - bln\theta \quad . \tag{10}$$

Comparing Eq. (10) with (5), we obtain

$$b = \hat{b}; \theta = exp(- \hat{a}/b) \quad .$$

The Weibull distribution parameters b and θ can be found through linear regressive parameters $\hat{a}$ and $\hat{b}$. Table 1 shows the results of applying linear regressive processing to the failure data of 24 machining centers.

The symbol No in the table denotes the machine number. Lines $1 - 18$ correspond to the fitted data of the machine numbers $1 - 18$, respectively. Line 19 corresponds to the fitted data of the other six machine tools.

From the data in the table we can see that $|R| > R_a$, so the effectiveness of linear correlation is very significant. That is to say, the failure distribution of MC fits the Weibull distribution. Meanwhile, the values of shape parameter b are all very near to 1, this means that this type of distribution perhaps fits the special case of Weibull distribution-the exponential distribution.

EXPONENTIAL DISTRIBUTION FITTING AND TESTING

The exponential, or strictly the negative exponential, distribution is probably the most widely known and used distribution in failure analysis and reliability evaluation. The most important factor for it to be applicable is that the hazard rate should be constant, in which case it is defined as the failure rate λ. The cumulative distribution function is given by.

Table 1   Linear regressive result
of Weibull distribution

No.	a	b	θ	R	R_a
1	-6.028	1.064	288.252	0.983	0.744
2	-7.715	1.334	112.379	0.989	0.591
3	-5.436	0.963	282.520	0.973	0.665
4	-6.001	1.146	187.824	0.976	0.591
5	-5.923	1.171	157.487	0.937	0.625
6	-6.614	1.306	158.426	0.988	0.591
7	-8.102	1.307	154.919	0.992	0.549
8	-6.211	1.212	168.143	0.980	0.591
9	-4.612	0.858	215.447	0.974	0.644
10	-7.051	1.424	141.503	0.974	0.576
11	-6.361	1.230	176.174	0.967	0.591
12	-5.373	1.038	176.680	0.962	0.625
13	-5.428	1.158	108.753	0.947	0.591
14	-7.057	1.503	109.420	0.968	0.625
15	-7.223	1.488	94.542	0.997	0.591
16	-7.813	1.404	97.974	0.994	0.688
17	-5.807	1.289	96.092	0.971	0.644
18	-5.787	1.120	175.445	0.968	0.625
19	-4.303	0.760	288.293	0.987	0.644

$$F(t) = 1 - exp(-\lambda t) \quad . \tag{11}$$

Rewrite Eq. (11) and take the logarithm, then

$$ln \frac{1}{1-F(t)} = \lambda t \quad . \tag{12}$$

We can evaluate $F(t_i)$ by Eq. (1), plot each couple $(t_i, ln \frac{1}{1-F(t_i)})$ in the rectangular coordinates, and fit out the line by linear regressive method. Fig. 3a shows the result of fitting the failure data of the above machine tool, Fig. 3b the other six machine tools. The linear regressive parameters are listed in Table 2.

a and b in Table 2 are parameters of linear regressive equation (5). Here b equals to λ of Equation (11), the meaning of R and R_a are similar to Table 1. Table 2 also reveals $|R| > R_a$, so that the effectiveness of linear correlation is very significant. That means the failure distribution of machining centers fits the exponential distribution with constant failure rate.

We test the goodness of fit based on the equipment reliability testing standard IEC 605 — 6[8]. One procedure of testing for the validity of a con-

stant failure rate assumption is the chi-square "goodness of fit" test. The test statistic is [7,8]

$$\chi^2 = 2\left(rlnT_0 - \sum_{i=1}^{r} lnT_i\right) \tag{13}$$

Table 2   Linear regressive parameters
of the exponential distribution

No.	a	b	R	R_a
1	0.035	0.003438	0.982	0.744
2	-0.433	0.014022	0.994	0.519
3	0.051	0.003532	0.950	0.665
4	0.058	0.005070	0.990	0.591
5	0.037	0.006293	0.973	0.625
6	-0.211	0.008219	0.990	0.591
7	-0.424	0.010107	0.992	0.549
8	-0.138	0.007178	0.979	0.591
9	0.236	0.003212	0.977	0.644
10	-0.028	0.007668	0.959	0.576
11	0.054	0.005489	0.978	0.591
12	0.109	0.004969	0.982	0.625
13	-0.430	0.015468	0.985	0.591
14	-0.099	0.010701	0.967	0.625
15	-0.482	0.017375	0.967	0.591
16	-0.558	0.017605	0.975	0.688
17	0.023	0.010565	0.973	0.644
18	0.110	0.005084	0.960	0.625
19	0.260	0.002223	0.981	0.644

where r is the total number of failure occuring in each equipment, T_0 is the total running time of the equipment, T_i is the cumulative running time up to the ith failure.

$$\chi_a^2(r) = \frac{1}{2}[Z_a + (2r-1)^{\frac{1}{2}}]^2 \quad . \tag{14}$$

Define $\chi_a^2(r)$ as the percentage point or value of the chi-square random variable with r degrees of freedom such that the probability that $\chi^2(r)$ exceeds this value is α. If we take the significance level $a = 0.05$, then $Z_a = Z_{0.05} = 1.64$; and $\alpha = 0.95$, $Z_{0.95} = -1.64$. Therefore, the critical values of the chi-square test for every machine tool can be found in accordance with failure number of the equipment from Eq. (14). The results of the goodness of fit test for the failure data of 24 machining centers are shown in Table 3.

From Table 3 we can see that the test statistics fall between the two percentage points of the chi-

91

square distribution, ie

$$\chi^2_{0.05}(r) < \chi^2 < \chi^2_{0.95}(r)$$

Table 3 Goodness of fit test

NO.	a	$\chi^2_{0.05}(r)$	$\chi^2(r)$	$\chi^2_{0.95}(r)$
1	0.10	15.1329	23.6181	38.5567
2	0.10	60.0652	91.5412	101.5244
3	0.10	23.0259	33.8133	50.6637
4	0.10	37.8805	66.0885	71.8091
5	0.10	46.3629	77.6998	83.3267
6	0.10	48.0741	51.4817	85.6154
7	0.10	48.0741	59.6680	85.6154
8	0.10	42.9545	62.9681	78.7351
9	0.10	34.5270	46.9196	67.1626
10	0.10	44.6563	63.5647	81.0333
11	0.10	41.2578	50.1967	76.4318
12	0.10	36.2006	47.5540	69.4890
13	0.10	70.6593	89.0132	115.0303
14	0.10	70.6593	87.9404	115.0303
15	0.10	81.2493	96.3102	128.4403
16	0.10	53.2333	64.3484	92.4564
17	0.10	67.1496	89.5306	110.5400
18	0.10	36.2006	63.5344	69.4890
19	0.10	19.8277	45.2132	45.8619

So we can say with confidence that the failure data of machining centers fit the exponential distribution.

CONCLUSION

From the above mentioned we can see that the failure distribution of machining centers fits the exponential distribution, that is to say, the failure rate of machining centers is constant. That provides the basis for estimating the reliability of machining centers. As is well know, the selection of reliability characterizatitions of equipments depends on the failure distribution. The mean time between failures (MTBF) is one of the widely-used characterizations, its mathematical definition is the expectation of the running time between failures.

If the constant failure rate is λ (see Eq. (11)), then $MTBF = 1/\lambda$, which can be estimated by

$$MTBF = \frac{1}{r} \sum_{i=1}^{n} t_i \quad . \tag{15}$$

where r is the sum of failure of all equipments tested; t_i is the actual running time of the ith equipment; n is the number of tested equipments. Eq. (15) can be used for estimating the reliability of machining centers only if their failure rate is constant, i. e. the failure time fits the exponential distribution.

REFERENCES

1. Jia Yazhou, Sense of urgency for enhancing the reliability of machine tools, Machine Tools, 1992 (4), P27—30 (in Chinese).

2. Jia Yazhou and Jia Zhixin, Fatigue load and reliability design of machine tool components, Int J Fatigue, 1993, 15(1), P47—52.

3. Ma Jian, Reliability research on machining centers, ME Thesis, Jilin University of Technology, P. R. China, 1994.

4. P. A. Tobias & D. C. Trindade, Applied Reliability, VNR Company Inc. 1986, P110—115.

5. K. C. Kapur & L. R. Lamberson, Reliability in Engineering Design, John Wiley & Sons, 1977, P220—260, 291—295.

6. R. E. Walpole & R. H. Myers, Probability and statistics for engineers and scientists, Macmillan Publishing Co. Inc., 1978, P280—308.

7. J. Neter, W. Wasserman & G. A. Whitmore, Applied statistics, Allyn and Bacon Inc., 1987, P409—418, 724.

8. IEC 605—6, Equipment reliability testing, part 6.

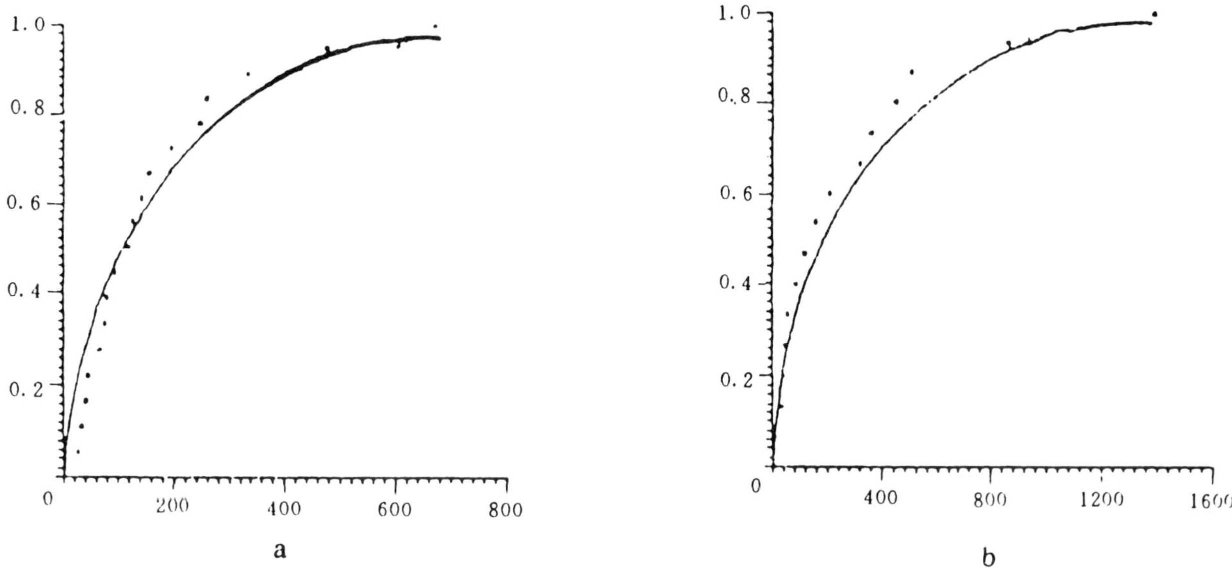

Fig. 1   distribution point diagram of field failure

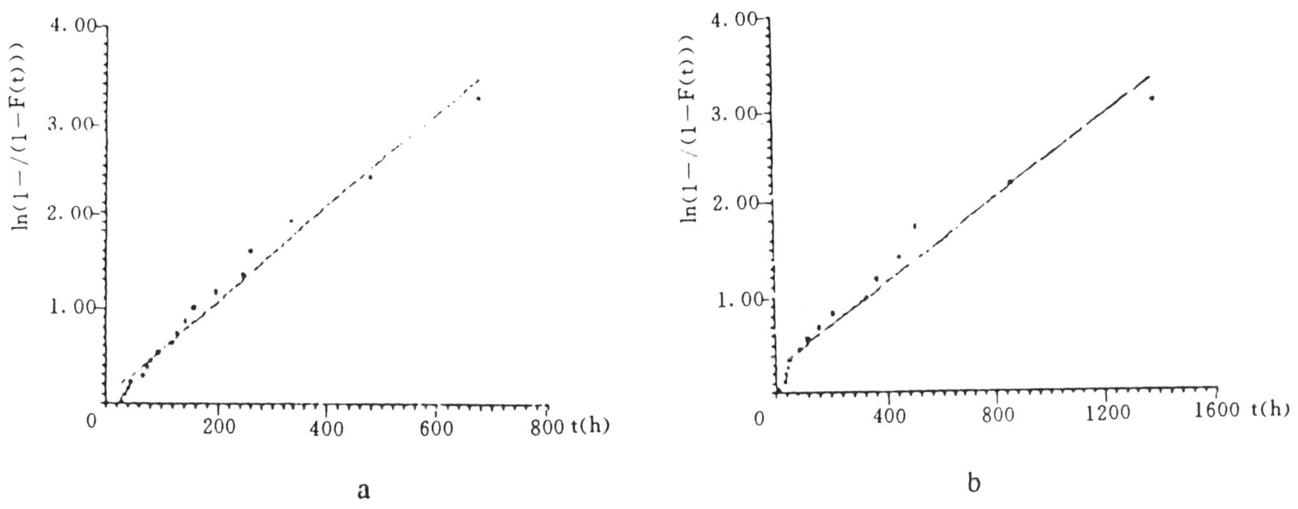

Fig. 3   Exponential distribution fit of failure data

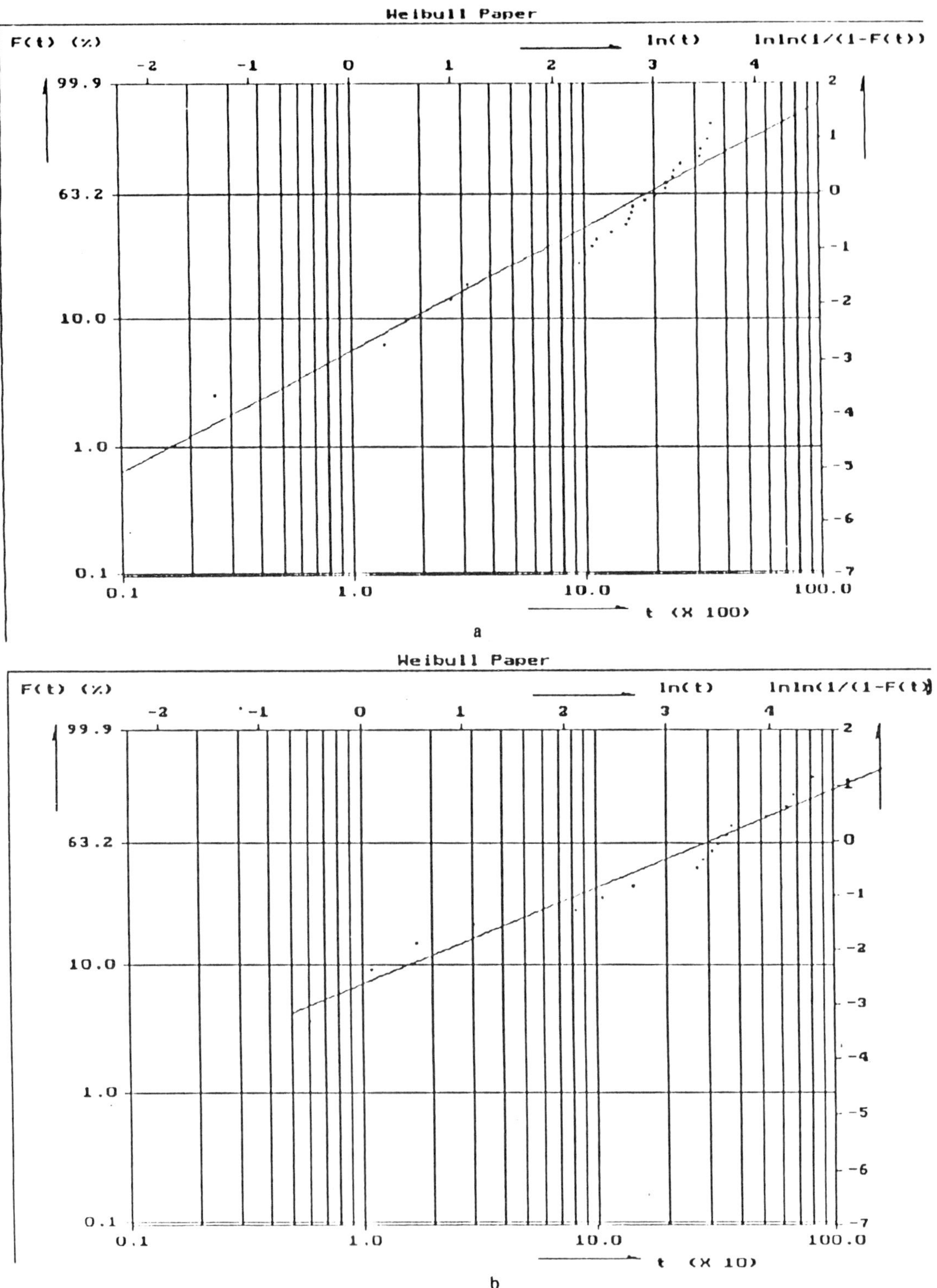

Fig. 2   Weibull distribution fit of failure data

RELIABILITY ANALYSIS OF MACHINING CENTERS (MC)

Jia Yazhou, Wang Guiqin, Ma Jian
Mechanical Engineering Department
Jilin University of Technology
Changchun 130025
P. R. China

Jia Zhixin
Mechanical Engineering Department
Harbin Institute of Technology
Harbih 150001
P. R. China

ABSTRACT

This paper reports on work which considered the reliability of 20 cutter-changeable CNC machine tools — machining centers in China. Results were gathered by a combination of questionnaire and personal interviews supported by several detailed case studies of individual manufacturing companies. The results indicate that the reliability weak links exist in the three subsystems and the tool magazine and manipulator are the key in increasing the reliability of machine tools.

Key words: reliability analysis, machining center, failure distribution, weak link

INTRODUCTION

Machining centers (MC) are high-grade computer numerically controlled (CNC) machine tools. Many characters such as high level of productivity, high accuracy and high automation make them an important mark of the industrial modernization. On the other hand, with MC function being extended and its complexity on the rise, the probability of machine tool failure is also increasing. Hence, the problem of MC reliability has gained importance [1]. But up till now, there has been little research on MC reliability. As may be imagined, frequent failures seriously hamper the broad-use of MC. Thus it is clear that failure analysis and prevention are of critical importance.

A modern manufacturing system will in almost all cases contain, if not totally but at least partially, CNC machine tools. The reliability of those machine tools is critical to the successful operation of any production system [2].

FAILURE DATA-BANK

In this paper 20 machining centers are surveyed in the process of working, and field failure data are collected over a period of two years.

As is well-known, the reliability data are the foundation in studying the reliability. But the collection and accumulation of reliability data is a long and difficult process. The MC need to be tested and investigated in the field. Since it is difficult to collect and accumulate the failure data a computerised reliability data bank has not been set up yet.

This paper initiates the study in MC reliability. The failure data bank established by the authors provides the basis and the frame for studying reliability of MC systematically. The special software program is developed in the DBASE Ⅲ related data bank management system, and is called " the people's data bank", in order to popularize it. The data bank is managed and operated with this special software system. The data-bank has two user aspects, one is used to calculate and the other is used to analyse.

The method employed in collecting the required data is to supply the operational, maintenance, and production personnel with failure data forms and ask that forms be completed as directed. Forms containing the raw data are returned for processing [4].

The failure data bank contains the following information: machine code, machine number, batch number, failure code, failure date, failure mode, failure position, work time, down time, and records of repair. The information contained in the user aspect for calculation is the records of related failures which conform only with specification, and they are applied to assess and compute the reliability characterizations. The user aspect for analysis are the records of all related failures and they are applied to analyse the failure.

The failure data bank provides the practical data and information for analysis and research on failure distribution of MC in this paper. And it offers basis and references for successively studying MC reliability.

SYSTEM FAILURE CASE STUDY

Machining centers are expensive and high utilization is essential to ensure an acceptable return on the investment made. Consequently, high reliability is a key factor in their design and service. In order to know the reliability level of MC at present and improve it rapidly, the statistical analysis of MC failures must be carried out with the field records. The results of frequency analysis of subsystems in these 20 machining centers over a period of 2 years are given in full in ref. 3, and the data in Table 1 indicate the frequencies of machine zone responsible for breakdown.

Here a MC consisted of three parts. As shown in Table 1, the number of breakdowns for each part is different. The breakdown composition of every part should be given a detailed classification if we want to ascertain the failure distribution more clearly. Table 2 gives the breakdown composition of the control system.

Table 1

Machine zone responsible for breakdown	Number of breakdowns	Percentage of number of breakdowns (%)
Machine body	898	72.83
Control system	59	4.79
Accessories	276	22.38
Total	1233	100

Table 2

Breakdown position	Number of breakdowns	Percentage (%)
CNC system	35	59.32
Servo unit	17	28.81
Power supply	7	11.87
Total	59	100

From Table 2 we can conclude that the position of the most frequent breakdown is the CNC system. According to our survey they are imported from different countries and their failure rate is different. So we suggest that adopting the CNC system of better quality is an efficient means of enhancing the reliability of control systems

The failure proportion of accessories is given in Table 3.

Table 3

Breakdown position	Number of breakdowns	Percentage (%)
Pneumatic system	12	4.35
Body shield	74	26.81
Cooling system	69	25.00
Hydraulic system	114	41.30
Lubricating system	5	1.81
Others	2	0.73
Total	276	100

It is evident from Table 3 that the breakdowns of hydraulic system cover 41.3% of the total breakdown of accessories and its failure rate is the highest. It indicates that the hydraulic system is the reliability weak link of MC. Hence, while the hydraulic components are selected, their reliability level must be considered carefully.

Table 4

Machine zone responsible for breakdowns	Number of breakdowns			Percentage (%)
	Vertical	Horizontal	Sub-total	
Spindle box	27	108	135	15.03
Tool magazine	44	103	147	16.37
Mechanical arm	187	203	390	43.43
Feed systems				
X axis	4	32	36	4.01
Y axis	12	37	49	5.46
Z axis	8	20	28	3.12
Work table	14	75	89	9.91
Electrical system	11	13	24	2.67
Total	307	591	898	100

As has been already pointed out from Table 1, the number of breakdowns of machine bodies is the greatest in machining centers. According to their spindle orientation, all machining centers may be classified into two large groups: the vertical and the horizontal. There is much difference in frequencies of machine zone and component responsible for breakdown between the two groups of MC. The results of analysing breakdowns of machine bodies in the 20 above mentioned machines are given in Table 4.

It is evident from Table 4 that the number of breakdowns in tool magazine and mechanical arm is the greatest in machine bodies. Recall from the previous section that MC is defined as the CNC machine tools which can change tools automatically. So the system of changing tools is one of the most important parts of MC, and its reliability has a great influence on the successful operation of machines.

In order to compare intuitively the failure frequency of different machine zones, the frequency histograms of failures of the vertical and the horizontal MC are shown in Figs 1 and 2 respectively. In the figures the parenthesised numbers designate the frequencies of failures and the data out of the brackets denote the numbers of breakdowns. We can see from these two figures that the tool magazine and mechanical arm are the weak links of reliability in MC. The frequency of failures of spindle boxes for the horizontal MC is also high.

FAILURE MODE ANALYSIS

In order to analyse the breakdowns of weak links of MC in detail, we must research the failure mode of tool magazines and mechanical arms from the field records. Failure model is a summary on the basis of analysing the phenomenon and reason of breakdowns.

9 vertical and 11 horizontal machining centers are investigated in the study. The results of analysing breakdowns of mechanical arms for the vertical and the horizontal MC are given in Figs. 3 and 4, respectively. In the two figures the parenthesised numbers designate the frequencies of breakdowns and the numerals out of the brackets denote the breakdown numbers of the corresponding failure mode.

From Figs 3 and 4 we can see that the main failure modes of mechanical arms are the following: tool falling out, tool seizing up, failure to actuate, out of position, leakage of oil, hand bending, arm sinking and others, For the vertical MC, tool falling out is a critical mode of breakdown. For the horizontal MC, the dangerous modes are: seizing up the tools, out of position, failure to actuate and tool falling out. These breakdowns often hinder automatic tool change in MC.

Fig. 5 shows the frequency histogram of failure modes of tool magazines in horizontal MC. The failure modes mainly are out of position, non-return-to zero, looseness of location pins, failure to actuate, leakage of oil and others.

CONCLUSIONS

This paper has three main conclusions: First, reliability weak links exist in the machining centers. Second, the mainbody of MC is the machine zone of frequently occuring breakdowns. Third, the tool magazine and the manipulator are the keys to increasing the reliability of MC.

If the high levels of productivity demanded of MC are to be achieved then users of machines must keep careful records of their operation and maintenance activities. Wide public availability of this information will enable improvements in reliability to be made and assessments of the correct level of preventive maintenance to be performed. Reliability can only be achieved by thoughtful design and careful manufacture [5]. It can only be improved by feedback on breakdowns in use and that feedback can only be provided if users of MC keep records and provide data on maintenance in the field.

REFERENCES

1. Jia Yazhou and Jia Zhixin, Fatigue load and reliability design of machine tool components, Int J Fatigue, 1993, 15(1), 47—52.

2. P. F. McGoldrick and H. Kulluk, Machine tool reliability — a critical factor in manufacturing systems, Reliability Engineering, 1986, 14, 205—221.

3. Shu Ze, Failure analysis and research on machining centers, ME, Thesis, Jilin University of Technology, China, 1995.

4. E. Balagurusamy, Reliability Engineering, Tata McGraw-Hill Publishing Company Ltd. New Delhi, 1984, 246—252.

5. K. C. Kapur and L. R. Lamberson, Reliability in Engineering Design John Wiley & Sons, 1977, 1—3.

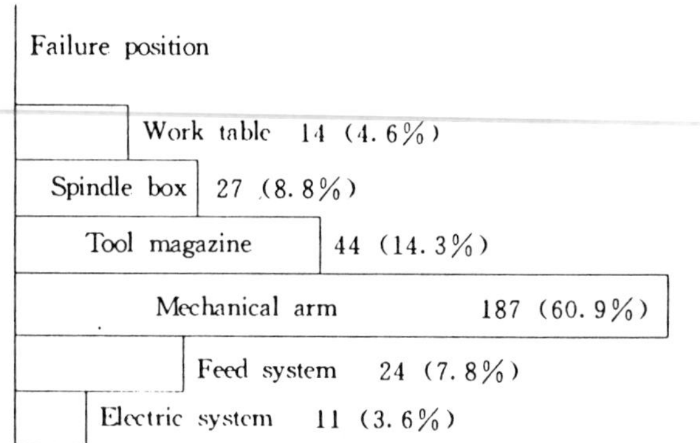

Fig. 1 Frequency histogram of failures for vertical MC

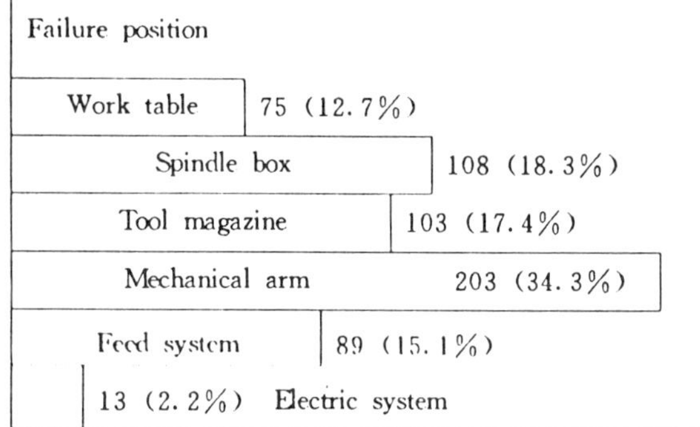

Fig. 2 Frequency histogram of failures for horizontal MC

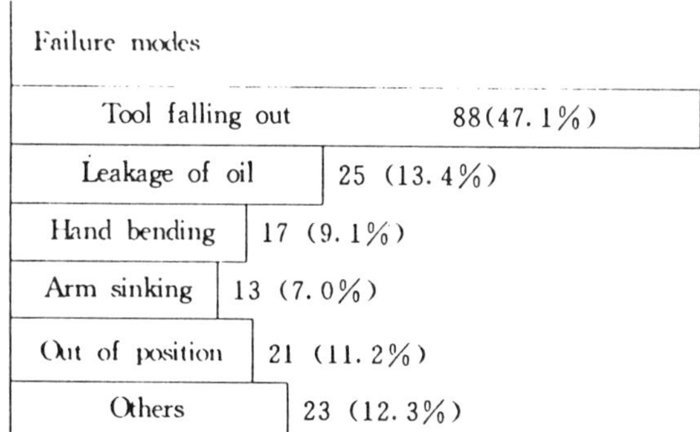

Fig. 3 Failure modes of mechanical arms for vertical MC

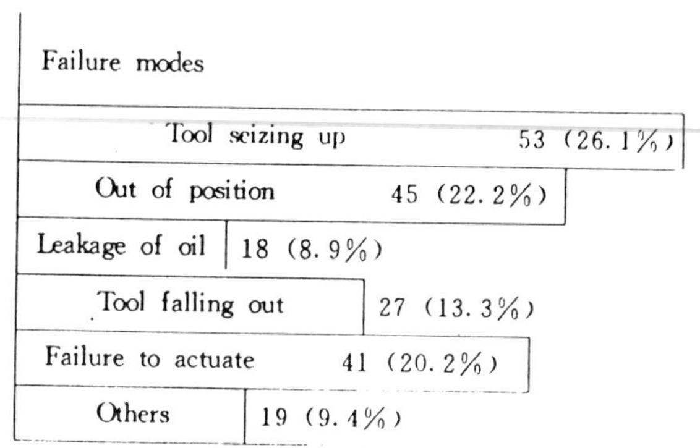

Fig. 4 Failure modes of mechanical arms for horizontal MC

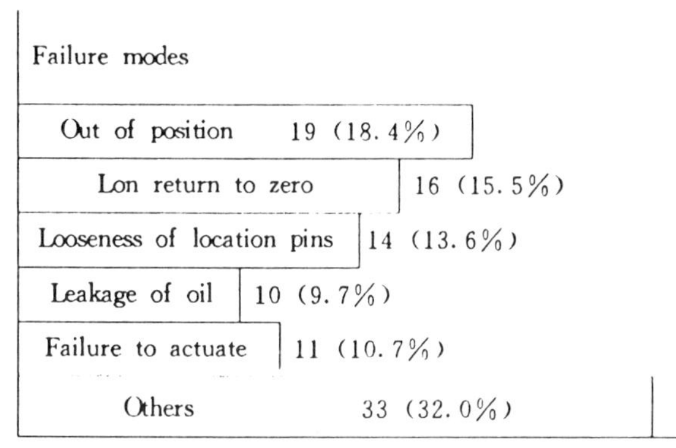

Fig. 5 Failure modes of tool magazines for horizontal MC

DE-Vol. 87, Reliability, Stress Analysis, and Failure Prevention
Issues in Emerging Technologies and Materials
ASME 1995

INVESTIGATION OF NONLINEAR BEHAVIOR
OF VEHICLE BODY PANELS UNDER SHIPPING CONDITIONS

Houchun Xia
Analytical Services Department
North American Operation Engineering Centers
General Motors Corporation
Troy, Michigan

David Y. Xue
Analytical Services & Materials, Inc.
Hampton, Virginia

ABSTRACT

The large deflection induced geometric nonlinear behavior of
vehicle body panels under shipping conditions is investigated in
this paper. The example structure presented is the door outer
panel of a passenger car. The studies include the nonlinear
finite element analysis for static displacements, stresses, and
normal modes, and the fracture mechanics analysis for fatigue
crack growth rate and fatigue life prediction involving panel
large deflections. The finite element code NASTRAN is
employed for the nonlinear analysis and the iteration strategy
used is the modified Newton-Raphson method. The fracture
mechanics analysis is based on the finite element method of
computing the energy release rate, G, using the virtual crack
closure technique. The stress intensity factor, K, is derived
from the G values, and the experimental-based Paris Law is
used for the panel fatigue life prediction. Although the
transverse deflection of the panel is large, i.e., close to the
panel thickness, at the container design loading levels, the
difference between the linear and nonlinear solutions is not
significant. One possible reason is that the midplane of the
panel is not stretched excessively due to the nature of the
constraints provided by the container dunnage supports on
panel edges. This situation exists for most vehicle body panels
under shipping conditions so that the linear analysis strategy
currently used in the automotive industry for the shipping
container design is feasible. The results of the fracture
mechanics analysis of the panel indicate that the dominant
failure mode for the assumed panel fatigue cracking is the
opening mode or Mode-I, and the linear analysis for the fatigue
life prediction of vehicle body panels under shipping conditions
provides reliable solutions. The geometric nonlinearities of the
panel in critical areas can be improved significantly by properly
designing the shipping containers.

INTRODUCTION

There has been a growing trend in the automotive industry to
ship sheet metal panels instead of assembled components from
panel stamping plants to vehicle assembly plants. Examples are
vehicle body side panels, door outer panels, roof panels, hood
outer panels, and deck lid outer panels. These panels are placed
into specially designed containers and shipped either by rail car or
by truck. Due to the complicated geometry of the panels and the
requirements for dimensional integrity and surface quality, these
panels are often individually supported in the container by several
dunnage at their edges. To obtain a proper design of the container
and prevent the panel from damage, appropriate structural analyses
are required in the container design process. The common
practice for the panel analysis involves linear static displacement
and stress analysis and linear normal mode analysis. For the panel
with the features which may cause stress concentration and fatigue
problems, linear fracture mechanics analysis is used. One inquiry
to the linear analysis strategy is raised from the large area-
thickness ratio of the panel. The typical thickness of the panel is
less than 1 mm. It is possible for these large, thin panels with
only edge supports to present the large deflection induced
geometric nonlinearities when subjected to the vertical and lateral
vibrations inherent in shipping environments. The purpose of this
paper is to investigate the nonlinear behavior of vehicle body
panels under shipping conditions and evaluate the validity of the
linear analysis currently employed in the shipping container design.

The example structure used for this study is a vehicle door outer
panel in truck shipping environments. The nonlinear structural
analyses for this panel, including static displacement and stress
analysis, normal mode analysis, and fracture mechanics analysis
involving large panel deflections, are presented in this paper and
the results compared to the corresponding linear solutions. While
the analytical methods of fracture mechanics analysis for linearly
deformed structures have been well established (Barsom and Rolfe,

1988 and 1987, Broek, 1986, Rybicki and Kanninen, 1977), the methods for evaluating the structures with geometric nonlinearities are still being actively investigated (Wang et al., 1995, Rankin et al., 1993, Starnes et al., 1994). The approach for the fracture mechanics analysis of the vehicle door outer panel adopted here, considering structural nonlinearities, is based on the finite element method of computing energy release rate, G, using the virtual crack closure technique (Wang et al., 1995, Wilkening, 1988). And then the Paris Law (Ewalds and Wanhill, 1984) is used to evaluate the fatigue crack growth rate and fatigue life of the panel under the prescribed shipping conditions.

FINITE ELEMENT MODEL

The vehicle door outer panel and the shipping container studied in this paper are illustrated in Figure 1. The multi-curved geometry of the panel has to be maintained during the panel shipment in the consideration of vehicle aerodynamics. The panel has small flanges at its four edges, see Figure 2, which will be bent (hammed) 90 degrees inward in order to hold the door inner panel to form a door assembly. To compensate for the curvatures of the panel and be able to obtain unwarped bending flanges, some notches are made in certain areas of the panel flanges. The root of the notches is a critical area where fatigue cracking should be prevented during panel shipment. The panels are supported on a bottom tray which is fixed on the container floor. The tray has many slots on its top surface to provide a support in both y and z directions. The two side dunnage bars shown in Figure 1 have finger-type spacers mounted on their inside surfaces as shown in Figure 3 in order to hold the panel and prevent it from moving in x and y directions. All of the dunnage supports provide only translational constraints to the panel and no constraints in any rotational directions. It can be seen from Figure 3 that the dunnage supports block the panel to move towards them but cannot keep the panel from moving away from them.

The finite element model of the door panel is built using a computer graphics program and then converted to a NASTRAN file for analysis. The model contains 1375 two-dimensional elements, CQUAD4 and CTRI3, as shown in Figure 4. The finite element mesh in the flange notch area where a microcrack is assumed has been refined for the panel fracture mechanics analysis, see Figure 5. The symbol s in Figure 4 indicates the nodal points where a single point constraint is applied to the model. The constraints represent the bottom, side, and top dunnage supports. The top hold-down bar, which is not shown in Figure 1, is located at the center of the top edges of the panels. This bar has similar finger-type spacers as that of the side dunnage bars and will prevent the panel from jumping in z direction as well as moving in y direction. Two loading conditions simulating truck shipping environments are considered. The first one is the acceleration in vertical direction which represents the vertical excitation of the truck to the panel, and the second one is the lateral acceleration corresponding to the lateral rolling of the truck during transportation.

EXAMINATION OF GEOMETRIC NONLINEARITY

The geometric nonlinear behavior of the door outer panel under the prescribed constraints and loading conditions are examined here. The finite element model discussed above is employed for this purpose. Unlike linear analysis in which solutions are obtained from a constant structural stiffness matrix directly, geometric nonlinear analysis has to be implemented using numerical iteration method in order to obtain a converged solution. In the nonlinear analysis the relationship between strain and displacement is nonlinear which depends on the magnitude of loadings. This means that the stiffness of the structure is no more a constant but a function of the loadings. To explore the nonlinear behavior of structures, the loads have to be divided into a number of increments and applied to the structure incrementally. For each load increment, several iterations may be needed to achieve convergence.

The relationship between strain and displacement for two-dimensional plates with large deflection induced geometric nonlinearities can be expressed as follows according to the von-Karman theory (Chu and Herrmann, 1956):

$$\{\epsilon\} = \{e_m\} + \{e_\theta\} + z\{\kappa\} \tag{1}$$

or

$$
\begin{Bmatrix} \epsilon_x \\ \epsilon_y \\ \gamma_{xy} \end{Bmatrix} = \begin{Bmatrix} u_{,x} \\ v_{,y} \\ u_{,y} + v_{,x} \end{Bmatrix} + \begin{Bmatrix} \frac{1}{2}w_{,x}^2 \\ \frac{1}{2}w_{,y}^2 \\ w_{,x}w_{,y} \end{Bmatrix} - z \begin{Bmatrix} w_{,xx} \\ w_{,yy} \\ 2w_{,xy} \end{Bmatrix} \tag{2}
$$

where the vectors $\{\epsilon\}$, $\{e_m\}$, $\{e_\theta\}$, and $\{\kappa\}$ are the total strain, infinitesimal linear strain, nonlinear strain due to the large rotation of the out-of plane displacement w, and the curvature of the deformed plate, respectively, z is the coordinate in the out-of plane direction, ϵ and γ are the normal and shear strain components and u and v the in-plane displacements in x and y directions, respectively, the comma denotes derivative operations, and the x and y in subscripts indicate the coordinates on which the operation works.

In finite element analysis, Equation (1) can be written as:

$$\{\epsilon\} = [B]\{u\} = ([B_L]+[B_N])\{u\} \tag{3}$$

where $\{u\}$ is the nodal displacement vector, and $[B]$ the element matrix which consists of the linear and nonlinear parts.

The stress and strain relation is linear for this study which obeys the Hooke's Law:

$$\{\sigma\} = [D]\{\epsilon\} = [D][B]\{u\} \tag{4}$$

where $\{\sigma\}$ is the element stress vector and $[D]$ the material matrix.

Based on Equations (3) and (4), the system equilibrium equation of the plates can be written as:

$$\{P\} = [K]\{u\} = ([K_L] + [K_N])\{u\} \tag{5}$$

with

$$[K_L] = \int_V [B_L]^T [D][B_L]\,dV \tag{6}$$

$$[K_N] = \int_V ([B_L]^T [D][B_N] + [B_N]^T [D][B_N] \\ + [B_N]^T [D][B_L])\,dV \tag{7}$$

where $\{P\}$ is the loading vector, $[K]$ the total stiffness matrix of the plate, $[K_L]$ the linear stiffness matrix, $[K_N]$ the nonlinear stiffness matrix due to the large rotation, the superscript T denotes the transposition of a matrix, and the integration is over the entire volume of the plates.

The iteration strategy employed in this paper is the modified Newton-Raphson method in which the equations to be solved at the ith step are (Noor and McComb, 1980):

$$[K_T]_i\{\Delta u\}_{i+1} = \{R\}_i \tag{8}$$

$$\{u\}_{i+1} = \{u\}_i + \{\Delta u\}_{i+1} \tag{9}$$

$$\begin{aligned}[K_T]_i &= \frac{\partial}{\partial\{u\}_i}(([K_L] + [K_N]_i)\{u\}_i) \\ &= [K_L] + \frac{\partial([K_N]_i\{u\}_i)}{\partial\{u\}_i}\end{aligned} \tag{10}$$

and

$$\{R\}_i = \{P\} - ([K_L] + [K_N]_i)\{u\}_i \tag{11}$$

where $[K_T]$ is the tangential stiffness matrix, $\{\Delta u\}$ the incremental displacement vector, and $\{R\}$ the unbalanced or residual nodal force vector which is the difference between the applied loads and the internal forces.

The iteration starts with an assumed displacement vector $\{u\}_0$, usually the linear displacement solutions. Substituting $\{u\}_0$ into Equations (3) through (7) and (10) and (11) yields $[K_T]_0$ and $\{R\}_0$. The values for $\{\Delta u\}_1$ and $\{u\}_1$ can then be obtained from (8) and (9). The iteration continues until the residual force $\{R\}$ becomes negligible, which is signified by a prescribed convergence criterion. The modified Newton-Raphson method is computationally efficient since the tangential stiffness matrix $[K_T]$ is evaluated only at the beginning of each load increment instead of for every iteration steps.

To ensure a reasonable accuracy for the nonlinear analysis of the door panel and consider computational costs, the applied loads are divided into ten equal increments for this study. After each load increment, a displaced element coordinate system is constructed for each element which follows and rotates with the element as the model deforms. The strategy for updating the element stiffness matrix adopted here is based on the solution convergent rates. At each load increment the number of iterations required to converge is estimated. The stiffness matrix of the panel is updated if (i) the estimated number of iterations exceeds a prescribed limit on the iterations for each load increment, (ii) the estimated time for convergence with current stiffness matrix exceeds the estimated time required for convergence with the updated stiffness matrix, and (iii) solution diverges. If the solution does not converge in the prescribed limit on the iteration numbers, the load increment will be bisected and the analysis repeated. The updated element stiffness matrix is expressed in the displaced element coordinate system and used to calculate the nonlinear displacements and stresses, and the quasi-nonlinear normal modes for the door outer panels. The quasi-nonlinear normal modes are computed using the following equation:

$$([K] + [K_N])\{\Phi\} = \lambda[M]\{\Phi\} \tag{12}$$

where λ and $\{\Phi\}$ are the eigenvalue and eigenvector to be solved, respectively, $[M]$ is the mass matrix of the panel, and $[K_N]$ evaluated using the nodal displacement vector $\{u\}$ but not using the eigenvector $\{\Phi\}$ through numerical iterations.

The finite element analysis for the panel is conducted on a CRAY-YMP supercomputer. Two static load cases are considered. The maximum load for Load-case I is the combination of 15g (fifteen times the force of gravity) in $-z$ direction and 7.5g in y direction, and that for Load-case II is 7.5g in $-y$ direction. The enlarged static deformations of the panel under the two load cases are illustrated in Figures 6 and 7, respectively. The undeformed shape of the panel is represented by the panel outlines as shown in the figures. The panel vibrates between the deformed shapes during transportation since the real load has a cyclic nature. It can be seen from the figures that the panel flange notch of interest is undergoing open and closed motions alternatively under the cyclic loadings. It is believed that this motion, i.e. the forced vibration of the panel, may cause the fatigue crack initiation and propagation at the root of the panel flange notch during shipment.

The linear and nonlinear finite element results of the displacement components in y and z directions for Node j under the two load cases are given in Figures 8 and 9, respectively, where the node j is located in front of the crack tip as shown in Figure 10. The y component, u_{jy}, represents the transverse deflection of the panel, and the z component, u_{jz}, represents the open and closed motion of the assumed microcrack. The maximum loadings of Load-cases I and II are divided into ten equal increments which form the horizontal axes of the figures. The results of von-mises stresses for element m, which is also shown in Figure 10, are plotted in Figure 11. From these results the following observations can be derived:

1) The nonlinear displacement and stress solutions of the panel under Load-case I are larger than the linear solutions, see Figures 8 and 11. At the maximum loading conditions studied here, the displacement components u_{jy} and u_{jz} and the von-mises stress σ_m are 7.7%, 12.7%, and 5.0% higher, respectively, when the geometric nonlinear behavior of the panel is considered. This means that the

deformation and stress conditions of the panel are more severe under the Load-case I when the large deflections of the panel are taken into account. This effect is more pronounced at higher loading levels. For Load-case II the nonlinear solutions for both displacements and stresses are slightly less than the linear ones, which indicates that the stiffness of the panel is actually increased when the panel deforms in -y direction. The relative error of the linear results compared to the nonlinear ones is below 3% at the maximum loading level in this case.

2) Under the loading levels currently used for the container design, i.e., 3g in vertical direction and 1.5g in lateral direction, the transverse deflection of the panel at the Node j, u_{jy}, is 0.649 mm which is close to the panel thickness, 0.75 mm. Therefore, the large deflection induced geometric nonlinearity of the panel should be assumed according to the theory of plates and shells (Timoshenko, 1959). The finite element results show, however, that the difference between the linear and nonlinear solutions at the design loading level, i.e., the second load step in the figures, is not significant. The maximum error for both linear displacements and stresses compared to the nonlinear solutions is below 1.8%, which is acceptable for most engineering applications. One possible reason for this is that the midplane of the panel is not stretched excessively due to the nature of the constraints provided by the dunnage supports to the panel edges. As mentioned above, the dunnage supports block the panel to move towards them but cannot keep the panel from moving away from them. Consequently, the membrane stiffening effect of the panel is not strong even when the panel is undergoing a large transverse deflection. It should be pointed out that the situation discussed here exists for most vehicle body panels during shipment. Therefore, the linear analysis strategy currently used in the automotive industry for the panel shipping container design is feasible according to this important observation.

3) The finite element results for the natural frequencies of the panel under the original container design are listed in Table 1. The quasi-nonlinear solutions are obtained from the panel deformed shapes at the design loading levels. It can be seen from the table that the maximum error of the linear analysis for the first four natural frequencies is 1.2% and the fundamental frequency of the panel is 29.353 Hz, which meets the dynamic design criterion for road shipping systems adopted in this paper to prevent the panel from resonance. The natural frequencies of the panel will vary as the panel deforms, for example, the fourth natural frequency of the panel will vary in a range of 1.261 Hz under the container design loading levels. This implies that the panel will resonate in a larger frequency range if the geometric nonlinear behavior of the panel is taken into account.

EVALUATION OF FRACTURE MECHANICS BEHAVIOR

<u>Computation of Energy Release Rate Based on Virtual Crack Closure Technique</u>

It is well known that the energy release rate, G, is an important parameter to predict structural integrity in the fracture mechanics analysis. This parameter represents the elastic energy per unit crack surface area which is available for infinitesimal crack extension. The method of calculating the energy release rate, G, employed in this paper is based on the virtual crack closure technique. According to this technique, the energy release rate corresponding to a small increment of the crack is equal to the energy required to close the crack by that increment. Therefore, the energy release rate, G, of a plate can be expressed as the limit of the increment of work done by the internal force of the plate to close the opened crack surface, ΔW, with respect to the corresponding crack closure increment, Δa, as Δa approaches zero (Wilkening, 1988), i.e.,

$$G = \frac{1}{t} \cdot \lim_{\Delta a \to 0} \frac{\Delta W}{\Delta a} \qquad (13)$$

where t is the thickness of the plate in the crack tip area.

The finite element implementation of Equation (13) can be written as:

$$G = \frac{1}{2t\delta} \sum_{k=1}^{6} F_{ik}(u_{jk} - u_{j'k}) \qquad (14)$$

where F_{ik} is the kth component of the nodal force at the crack tip point i, see Figure 7, u_{jk} and $u_{j'k}$ are the kth components of the nodal displacements at grid points j and j', respectively, and δ is a small increment of the crack closure. The nodal force F_{ik} is the summation of the forces contributed by the elements m and n to the node i. The grid points j and j' are coincident before the crack propagates the increment δ. All the three translational and three rotational components of the nodal forces and relative displacements are taken into account in Equation (14). Substituting the finite element results of the nodal forces and displacements into (14), the corresponding values for the energy release rate, G, can be obtained. It should be noticed that the relationship between the nodal forces and the corresponding nodal displacements can be assumed linear if the crack closure increment, δ, is sufficiently small.

The finite element model discussed above, which includes a small crack at the root of the panel flange notch as shown in Figure 4, is used here to compute the energy release rate for the door outer panel under the prescribed shipping conditions. The crack length, a, of the model is assumed to be 0.61 mm and the crack increment, δ, be 0.1 mm. The two load-cases mentioned above are considered for this study. The Load-case I is assumed to be the worst case for the panel crack initiation and propagation at the flange notch root of interest, while Load-case II represents a loading in the opposite direction to that in Load-case I.

The linear finite element results of the nodal forces and displacements and the resultant G values obtained from Equation (14) are listed in Table 2. It can be seen that the energy release rate of the panel corresponding to the nodal force in z direction, i.e., G_3, is dominant among the six components of the total energy release rate, G. This force component is in the crack open and closing direction which is referred to as Mode-I loading in the fracture mechanics. For both Load-cases I and II, 99.2% of the

Mode	Linear solution (Hz)	Quasi-nonlinear solution					Δf (Hz)
		Load-case I		Load-case II			
		f_I (Hz)	$e_{f\,I}$ (%)	f_{II} (Hz)	$e_{f\,II}$ (%)		
1	29.353	29.406	0.2	29.160	0.7		0.246
2	47.299	47.554	0.5	47.001	0.6		0.553
3	59.389	59.187	0.3	58.514	0.2		0.327
4	65.615	66.431	1.2	65.170	0.7		1.261

Nodes: $e_{f\,i}$ - Relative error of linear analysis for Load-case i;
Δf - Natural frequency range of the panel considering geometric nonlinearities.

TABLE 2   LINEAR SOLUTIONS UNDER THE ORIGINAL CONTAINER DESIGN

k	Load-case I				Load-case II			
	F_{ik} (N)	$u_{jk} - u_{j'k}$ (cm)	G_k (N/cm)	P_k (%)	F_{ik} (N)	$u_{jk} - u_{j'k}$ (cm)	G_k (N/cm)	P_k (%)
1	0.0763	0.0000	0.0001	-	0.0606	0.0000	0.0000	-
2	4.1900	0.0000	0.0291	0.40	1.5300	0.0000	0.0040	0.50
3	56.400	0.0002	6.2800	99.2	21.000	0.0001	0.8600	99.2
4	0.0838	0.0004	0.0206	0.30	0.0302	0.0001	0.0030	0.30
5	0.0011	0.0050	0.0036	0.05	0.0004	0.0015	0.0000	-
6	0.0108	0.005	0.0036	0.05	0.0041	0.0002	0.0000	-

Notes: G_k - Energy release rate corresponding to the index k;
P_k - Energy release rate percentage corresponding to the index k, $P_k = G_k / G$;
Refer to Equation (14) for the definitions of other symbols.

total energy release rate of the panel contributes to the Mode-I energy release rate. Therefore, the dominant failure mode for the panel fatigue cracking is Mode-I or the opening mode. In the following studies only this failure mode is taken into consideration.

The procedure of computing the energy release rate, G, for the nonlinear analysis adopted in this paper includes four steps: (i) Computing the nonlinear nodal displacements and element forces in the crack tip area using NASTRAN solution sequence 106; (ii) Comparing the linear and nonlinear results of the element forces to obtain the ratios for the element forces between the linear and nonlinear solutions at different loading levels; (iii) Using these ratios and the linear solutions for nodal forces to calculate the nonlinear nodal forces; (iv) Substituting the nonlinear displacements and nodal forces into Equation (14) to solve for the nonlinear energy release rates.

The results of the Mode-I energy release rate for the vehicle door outer panel under the prescribed shipping conditions for the original container design are plotted in Figure 9. The relative error of the linear solutions compared to the nonlinear ones at the maximum loading levels studied here is 7.9% for Load-case I and 6.1% for Load-case II. Again, at the container design loading level, i.e., the second load step in the figure, the results of the linear and nonlinear solutions are very close.

Fatigue Life Prediction

The experiment-based Paris Law is used here to evaluate the fatigue crack growth rate of the vehicle door outer panel under shipping conditions. This law can be expressed as (Ewalds and Wanhill, 1984):

$$\frac{da}{dn} = C\,(\Delta K)^m \tag{15}$$

where a is the crack length, n the number of loading cycles, $\Delta K = K_{max} - K_{min}$ the stress intensity factor range in the vicinity of the crack tip, K_{max} and K_{min} are the stress intensity factors corresponding to the maximum and minimum loading conditions, respectively, and C and m the material constants which can be determined from material experiments.

The K values required to solve Equation (15) can be determined based on the relationship between the energy release rate, G, and the stress intensity factor, K. For tensile loading and plane stress condition, this relationship can be written as:

$$K = [EG]^{1/2} \tag{16}$$

where E is the elastic modulus of the material.

Assuming that the Load-case I is corresponding to the maximum loading condition for the panel crack initiation and propagation and Load-case II to the minimum loading, the values for K_{max} and K_{min} can be computed by substituting the resultant G values into (16). The material constants C and m in (15) for low-carbon structural steel are 5.01×10^{-12} and 3.1, respectively, providing that the unit for a is meter and for ΔK is $\mathrm{MPa} \cdot \mathrm{m}^{1/2}$ (Ewalds and Wanhill, 1984). Substituting these values into (15) yields the fatigue crack growth rate, da/dn, for the door outer panel under the prescribed shipping conditions. The results are listed in Table 3. Furthermore, the relationship between the crack length, a, and the loading cycle, n, can be derived by integrating Equation (15). This relation can be used either to predict the permitted loading cycles for an allowable maximum crack propagation length or to estimate the crack growth length according to a given loading cycles. The allowable loading cycles for the assumed notch crack of the door panel to grow 1 mm are evaluated here. The result obtained from the nonlinear analysis is 4.272×10^4 and that from the linear analysis is 4.193×10^4. The difference between the two results is 1.8%, and the linear analysis leads to a more conservative solution. This shows that the linear fracture mechanics analysis for the fatigue life prediction of the vehicle body panels under the shipping conditions can provide reliable solutions.

EFFECT OF DESIGN MODIFICATION

The fatigue crack growth rate of the vehicle door outer panel with the original container design obtained in the last section is not acceptable for the assumed truck shipping distance of about 200 miles. To improve the fracture mechanics behavior of the panel and extend the panel fatigue life, the original container design has been modified to achieve a reduction in the vibration and stress intensity of the panel around the flange notch area of interest. The modification involves lowering the original side dunnage bars towards the critical area and adding another dunnage bar on each side of the container to hold the panel at the edges below the notch of interest. The finite element model of the panel shown in Figure 4 is modified accordingly by repositioning and adding the single point constraints to the panel side flanges at the corresponding locations. The static deformations of the modified model under the Load-cases I and II show that the deflections of the panel flange notch area of interest are reduced significantly. The linear and nonlinear solutions for the displacements of Node j in y and z directions are plotted in Figure 13. It can be seen that the geometric nonlinearity of the panel around the critical area is improved with the modified container design. The displacement of Node j in y direction, u_{jy}, representing the transverse deflection of the panel, is reduced from 0.649 mm to 0.00252 mm, which is orders less than the panel thickness, at the container design loading levels. Therefore, it can be assumed that the panel will behave linearly under the modified container design. The comparisons of some linear analysis results between the different container design levels under the Load-case I are given in Table 4. It can be seen that the nodal displacements and forces and the energy release rate of the panel are reduced significantly with the modified container design, while the allowable loading cycles for the panel crack to grow 1 mm is increased by 55 times, which means that the allowable shipping distance will be increased by 55 times.

CONCLUSIONS

The large deflection induced geometric nonlinear behavior of vehicle body panels under the shipping condition is investigated in this paper. The example structure presented is the door outer panel of a passenger car. More severe deformation and stress conditions of the panel are obtained from the nonlinear finite element analysis compared to the linear ones. Although the transverse deflection of the panel is large, i.e., close to the panel thickness, at the container design loading levels, the difference between the linear and nonlinear solutions is not significant. The maximum error of the displacements, stresses, and natural frequencies is below 1.8%, which is acceptable for most engineering applications. One possible reason is that the midplane of the panel is not stretched excessively since the dunnage support at the panel edges cannot keep the panel from moving away from them. This situation exists for most vehicle body panels under shipment so that the linear analysis strategy currently used in the automotive industry for the shipping container design is feasible. The results of the fracture mechanics analysis of the panel indicate that the dominant failure mode for the assumed panel fatigue cracking is the opening mode or Mode-I, and the linear analysis for the fatigue life prediction of the vehicle body panels under shipping conditions can provide reliable solutions. The geometric nonlinearities of the panel in the critical area are improved significantly by the proposed modifications for the original shipping container design.

REFERENCES

Barsom, J.M., and Rolfe, S.T., 1988, "Fracture Mechanics in Failure Analysis," *Fracture Mechanics: Eighteenth Symposium, ASTM STP 945*, D.T. Read and R.P. Reed, Eds., pp. 443-467.

TABLE 3   COMPARISON OF THE LINEAR AND NONLINEAR RESULTS FOR THE ORIGINAL CONTAINER DESIGN

Quantity	G_{I1}	G_{I2}	K_{max}	K_{min}	da/dn	N
Linear solution	6.279	0.862	11,206	-4,152	2.385×10^{-8}	4.193×10^{4}
Nonlinear solution	6.297	0.818	11,222	-4,045	2.341×10^{-8}	4.272×10^{4}
Relative error (%)	0.3	5.4	0.1	2.6	1.9	1.8

Notes: Units: G - N/cm, K - N · cm$^{-3/2}$, da/dn - m/cycle;
G_{Ii} = Mode-I energy release rate under Load-case i, i = 1, 2;
N = Prediction of loading cycles corresponding to the crack growing 1 mm.

TABLE 4   COMPARISONS OF THE DIFFERENT CONTAINER DESIGN LEVELS UNDER LOAD-CASE I

Quantity	u_{jy}	u_{jz}	$u_{jz} - u_{j'z}$	F_{iz}	G	N
Original design	6.49×10^{-2}	4.60×10^{-2}	1.67×10^{-4}	56.44	6.329	4.160×10^{4}
Modified design	2.52×10^{-4}	6.24×10^{-3}	4.31×10^{-5}	15.10	0.477	2.267×10^{6}
Modified/original	0.004	0.136	0.258	0.268	0.075	54.495

Notes: Units: u - cm, F - N, G - N/cm;
G = Total energy release rate of the panel;
N = Same as in Table 3.

Barsom, J.M., and Rolfe, S.T., 1987, *Fracture and Fatigue Control in Structures - Applications to Fracture Mechanics*, Prentice-Hall Inc., Englewood Cliffs, NJ.

Broek, D., 1986, *Elementary Engineering Fracture Mechanics*, 4th Ed., Martinus Nijhoff Publishers.

Chu, H.N., and Herrmann, G., 1956, "Influence of Large Amplitudes on Free Flexural Vibration of Rectangular Elastic Plates," *Journal of Applied Mechanics*, Vol.23, pp. 532-540.

Ewalds, H.L., and Wanhill, R.J.H., 1984, *Fracture Mechanics*, Edward Arnold and Delftse Uitgevers Maatschappij.

Noor, A. K. and McComb, H. G., Editors, 1980, *Computational Methods in Nonlinear Structural and Solid Mechanics*, Pergamon Press.

Rankin, C.C., Brogan, F.A., and Riks, E., 1993, "Some Computational Tools for The Analysis of Through Cracks in Stiffened Fuselage Shells," *Computational Mechanics*, Vol.13, pp. 143-156.

Rybicki, E.F. and Kanninen, M.F., 1977, "A Finite Element Calculation of Stress Intensity Factors by a Modified Crack Closure Integral," *Engineering Fracture Mechanics*, Vol.9, pp. 931-938.

Starnes, J.H., Jr., Britt, V.O., Young, R.D., Rankin, C.C., Shore, C.P., and Bains, N.J.C., 1994, "Nonlinear Analysis of Damaged Stiffened Fuselage Shells Subjected to Combined Loads," *FAA/NASA International Symposium on Advanced Structural Integrity Methods for Airframe Durability and Damage Tolerance*, NASA CP-3274, Part 2, pp. 1045-1075.

Timoshenko, S., 1959, *Theory of Plates and Shells*, 2nd Edition, McGraw-Hill.

Wang, J.T., Xue, D.Y., Sleight, D.W., and Housner, J.M., 1995, "Computation of Energy Release Rates for Cracked Composite Panels with Nonlinear Deformation," *AIAA Paper 95-1463*.

Wilkening, W.W., 1988, "Computation of the Crack-Tip Energy Release Rate for Cyclic Crack Growth," *Fracture Mechanics: Eighteenth Symposium, ASTM STP 945*, pp. 1070-1082.

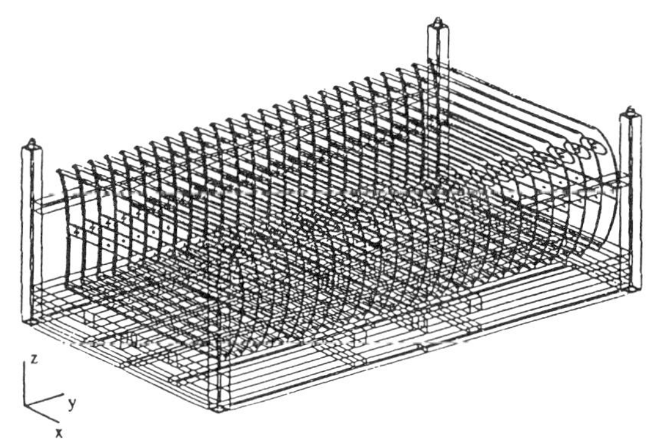

FIGURE 1   PANELS IN THE CONTAINER

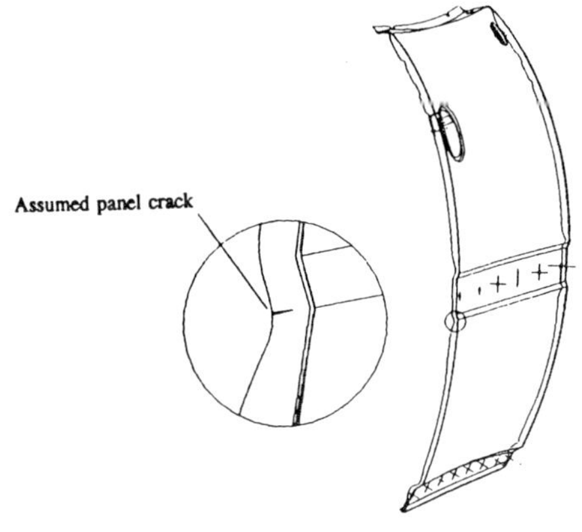

FIGURE 2   PANEL FLANGES AND NOTCHES

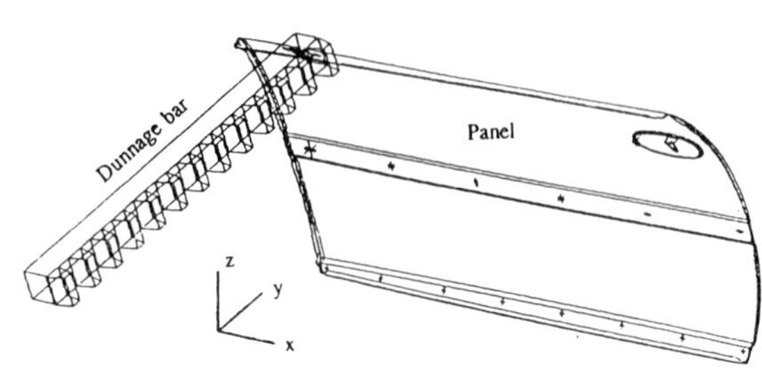

FIGURE 3   THE VEHICLE DOOR OUTER PANEL

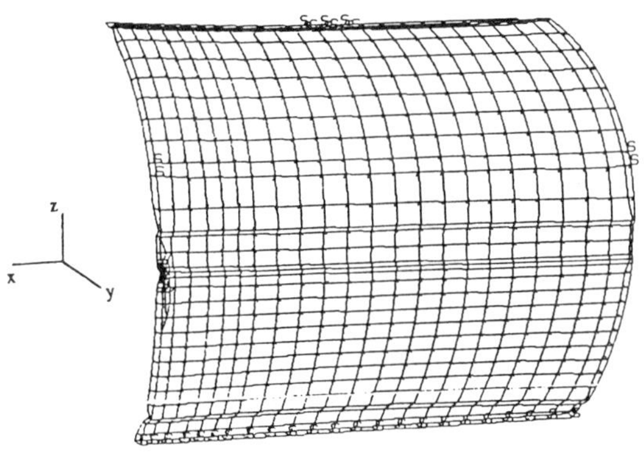

FIGURE 4   FINITE ELEMENT MODEL FOR THE
DOOR PANEL

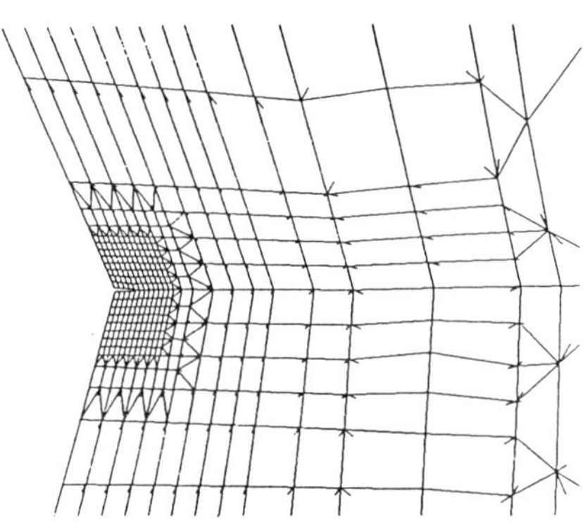

FIGURE 5   FINE MESH IN THE CRACK AREA

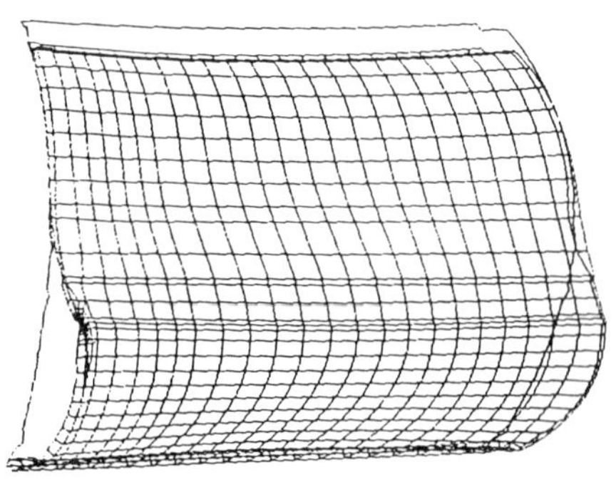

FIGURE 6   PANEL DEFORMATION, ORIGINAL
CONTAINER, SUBCASE 1

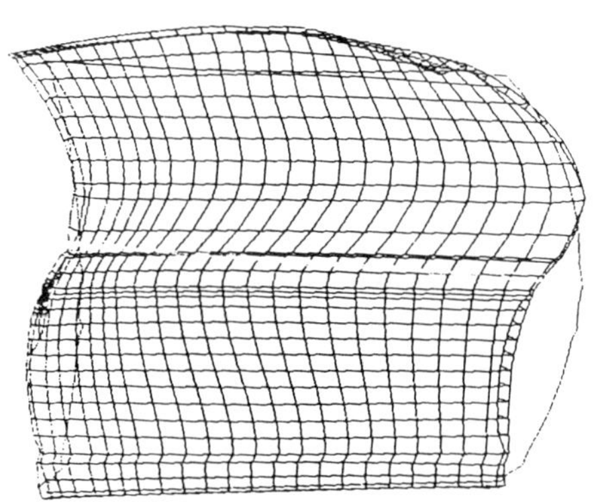

FIGURE 7   PANEL DEFORMATION, ORIGINAL
CONTAINER, SUBCASE 2

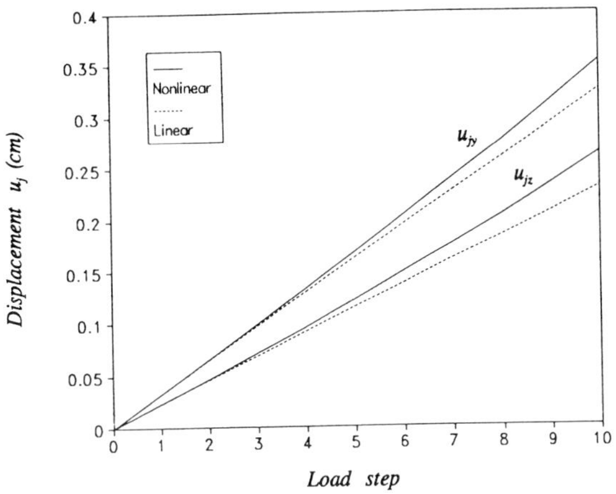

FIGURE 8   DISPLACEMENT OF NODE J, LOAD-CASE I

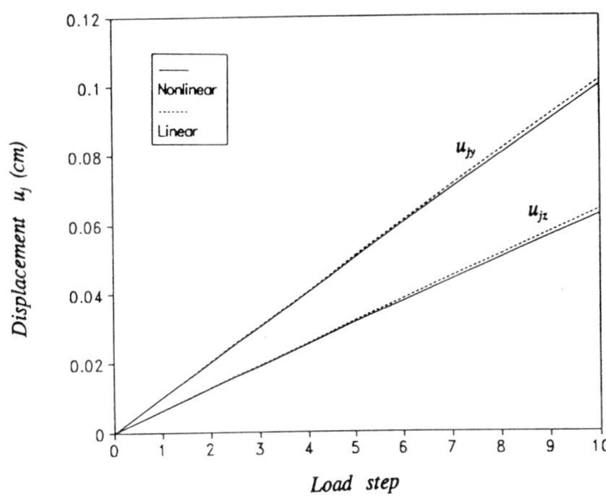

FIGURE 9   DISPLACEMENT OF NODE J, LOAD-CASE II

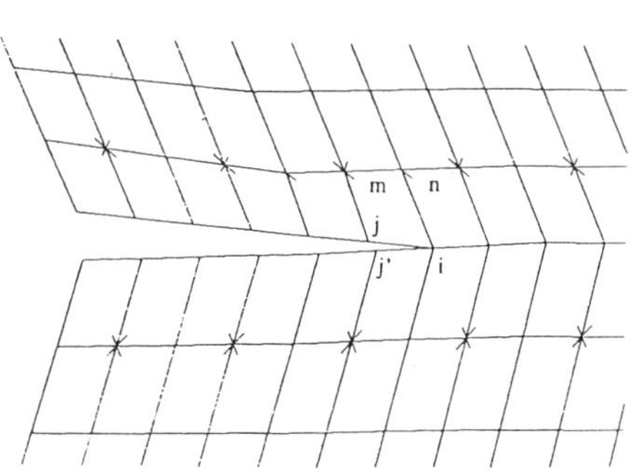

FIGURE 10   FINITE ELEMENT MESH IN CRACK TIP AREA

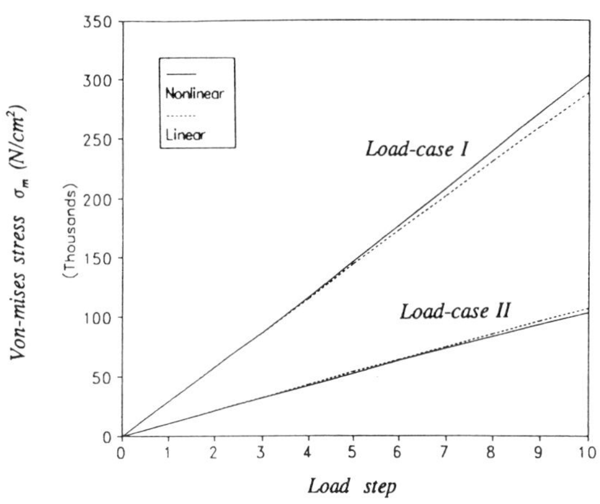

FIGURE 11   VON-MISES STRESSES OF ELEMENT M

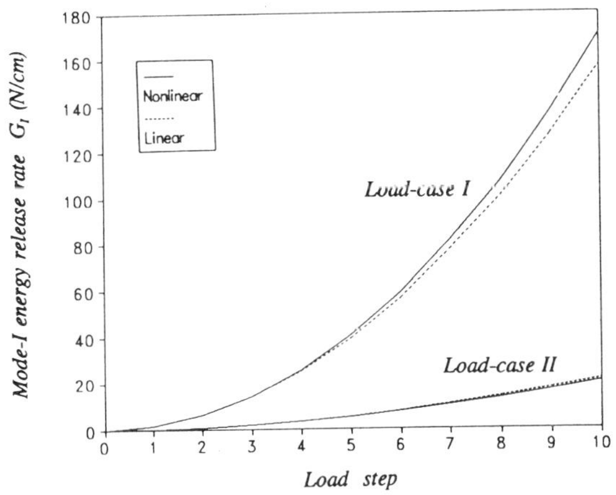

FIGURE 12   ENERGY RELEASE RATE OF THE PANEL

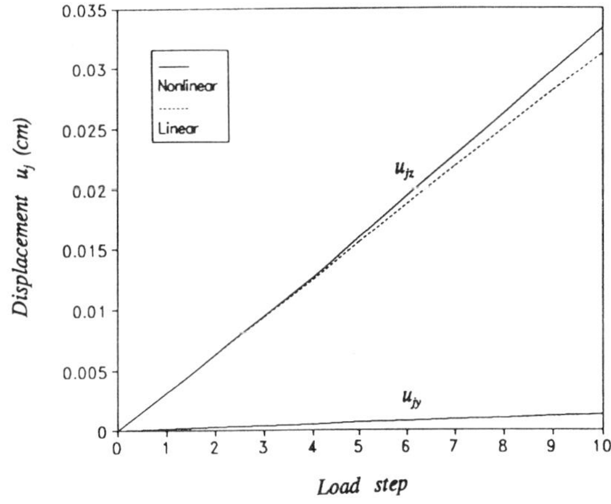

FIGURE 13   DISPLACEMENT OF NODE J WITH MODIFIED CONTAINER DESIGN

DE-Vol. 87, Reliability, Stress Analysis, and Failure Prevention
Issues in Emerging Technologies and Materials
ASME 1995

SIMULATION OF THE DROP TEST OF A PACKAGED REFRIGERATOR

Seoggwan Kim, Wonjoon Cho
LG Production Engineering Research Center
Pyungtaek
Korea

L.T. Kisielewicz, A. Petitjean, K. Matsumura
ESI Asia
Tokyo
Japan

ABSTRACT

The paper presents a feasibility study for carrying out numerical simulation of packed appliances. The needs of industrial companies in that field are discussed and some typical examples are briefly introduced. A more detailed presentation of the simulation of a lateral side drop test of a packed refrigerator is provided and some recommendations drafted for further work to make that solution more effective for the industrial companies.

INTRODUCTION

Packing of artifacts raises two classes of problems to manufacturers. The first class is to harden packings and artifacts in order to reduce the losses due to mishandling generated shocks and impacts. The second class of problems is related to the implementation of incoming regulations for protecting the environment and their influence on the industrial packing habits.

Hardening packings is aiming at reducing the losses due to mishandling before delivery to the consumers. Some major home appliance manufacturers claim losses of about 3% of their gross production during handling before delivery. While this figure looks rather small, when multiplied by the volume of production, it results in considerable amount of revenues.

The target of hardening packings is to reduce the amount of these losses either by reducing the number of artifacts damages during shocks and impacts or at least by protecting the main components so that they can be retrieved and reused.

In order to achieve these goals, the packings must absorb the kinetic energy without overstraining or overshaking the artifact. The criteria for overstrain and overshake are artifact dependent and cannot thus be defined as a universal value.

As of today the most popular shock absorbing material used in packings is expanded polystyrene (EPS). This situation will change in the coming years due to some guidelines and regulations aimed at the protection of the environment. It is expected that the disposal of used EPS in trash deposit will be banned because EPS is not biodegradable and that all EPS used by industry will have to be reprocessed because disposing of it by burning produces hazardous gases.

Considering this situation, a number of manufacturers worldwide have initiated R&D programs to find alternate shock absorbing materials. The number of solutions investigated is quite large and no single solution has yet emerged as the best one.

The procedure usually used in the R&D programs is to rely on physical tests for assessing the energy absorption capacity. Compared to other situations where energy absorption is a major design criteria (vehicle crashworthiness for example), these tests are rather inexpensive and could be multiplied without overburdening the R&D budgets. However, these tests provide limited insight into the structural behavior during shock application. An optimization procedure thus requires a very large number of trial-and-error tests which will eventually represent a significant cost and also lengthen the design cycle.

The paper presents an alternate approach to physical testing. Numerical simulation techniques have been developed and validated many years ago for simulating crashes of vehicles. These techniques initially limited to ductile steel structures have evolved over the years to cover more and more material types including ductile-brittle (aluminum), brittle (glass, plastic, composite, etc.), foams (urethane, polyurethane, etc.). These capabilities provide all the means needed to carry out simulation of drop tests of packed artifacts.

A feasibility project was carried out to demonstrate the possibility of using numerical simulation as an alternate to physical tests. The artifact selected was a 400 liters refrigerator in its traditional packing including EPS pads, wood, cardboard, and nylon straps.

The paper describes the numerical method used, the modeling assumptions especially for the packing materials and the results obtained including comparisons with test results. Conclusions about the needs for applying numerical technology are drawn and axes for further studies are outlined.

PHENOMENOLOGY

The consumer and home appliances and their packing are complex structures including a wide variety of material and components which do not behave the same way during a shock loading. Usually, the body structure is made of metal, a material yielding under excessive strains, and/or plastic, a stiff brittle material accumulating damages before it suddenly breaks, and/or rubber like material with an elastic incompressible behavior.

The appliances also include electronic parts which are sensitive to high frequency vibrations requesting shock spectra or power density spectra for separate equipment qualification and criteria verification, and mechanical parts like engines, gears, transmissions which are sensitive to structural deformations. The optical parts made of glass or plastic are sensitive to friction induced wear and should not suffer permanent deformation should they remain in proper state after the shock. Joints, like glue, welding, or riveting, represent local discontinuities in the geometry, and in the material properties of the appliances. Joints are prone to concentrate stresses if they would be structurally stiffer than the joined parts. Eventually, the packing of appliances are of interest. These can be made of various materials such as paper, cardboard, wood, foam, rubber, etc.

Altogether, the consumer and home appliances exhibit a variety of shapes and materials in rather equivalent proportions all calling for numerical modeling techniques to carry out shock simulation.

The shock of an unpacked falling appliance on the ground is a very short phenomena, especially when the ground surface is hard. Typically, such phenomenon lasts for a few milliseconds. The main phenomenon is a wave propagation involving damage accumulation and fracture. For shock on soft ground surfaces, such as carpeted grounds, rebounds and multiple shocks can happen, including some mechanical or geometric changes during the flight like the opening of a case.

When the appliance is packed, the impact can be longer as the kinetic energy is absorbed through mechanical deformations of the pack or some parts of it specially designed for that purpose.

The survivability of the appliance cannot be measured only by apparent structural integrity. There are also needs to assess the survivability of critical components to the high frequency vibrations involved as well as the damage accumulation in different parts.

PHYSICAL DROP TESTS

With a growing number of advanced appliances becoming integral part of everyday usage, such as laptop computers, a need has emerged to assess the response of these devices to shock due to drops and free falls.

Consumers request more robust appliances and more personalized ones. They want appliances which can survive some handling incidents, mainly free falls on the ground, and they do not want to use everybody else's model. This pressures the manufacturers to offer a wide variety of more robust appliances.

Also the fierce competition among them calls for shorter design cycles as the first product to hit the market is likely to grab the major market share.

At the other end of the range, home appliances are usually large and heavy and thus do not risk any mishandling by the consumers. However, some major manufacturers recognize losing about 4% of their gross production due to shocks occurring when the packed appliances are handled in storehouses or during transportation for delivery at shops or at customer houses. Reducing these losses up to 50% represents a major goal for the manufacturers. This goal can be reached by hardening the packing and the appliance in order to make them more shock resistant. Another target would be to protect the major equipments so that in case of shock, even if the body is damaged beyond repair, these equipments (representing a major share of the production cost) can be retrieved and reused.

While there is no general regulation applicable to crashworthiness of packings, in some sector of industry, regulations indirectly address that issue (Kadoya, 1990). Food is a perishable product sensitive to aggression like oxidation and packing is thus a necessity for conservation. Some regulation require that the packing of food be shock resistant, i.e. that for some intensity of shock, the packing won't rupture and endanger the food. In other sectors of the industry, while there is no regulation, the practice is to carry out drop tests on prototype series of the artifacts to optimize their design and that of their packing. Such is the case among the appliance manufacturers. As have been shown by some surveys, over 80% of appliance manufacturers in Japan do carry out such drop tests and the 20% remaining are considering starting similar procedures.

While the unit cost of drop tests on appliances is not very high, they cannot be carried out before prototypes are built, thus straining the design cycle duration. Also these tests do not provide very detailed insight into the behavior of the appliance during the shock. Due to the increasing miniaturization of the appliances, it is very difficult to place measuring devices inside them in order to monitor the stress and acceleration parameters in the critical components. Therefore the drop tests carried out by the manufacturer often results in one severity type criterion, i.e. providing figure like X% chances the appliance will or will not work properly after that type of free fall.

Hardening the packing of appliances is carried out empirically using some limited testing and resulting from statistical results from actual losses. The tedious trial-and-error approach provides only a very slow progress towards the optimized solution. The pressure in that field is also growing as some regulations are being issued restricting the usage of some of the most popular packing materials like expanded polystyrene pads (EPS) in packing. Replacement solutions are needed and no obvious answer is available.

SIMULATED DROP TESTS

1. Advantages of Computer Aided Engineering (CAE)

In the field of drop tests, numerical simulations can be used in the design process to achieve models where critical electronics

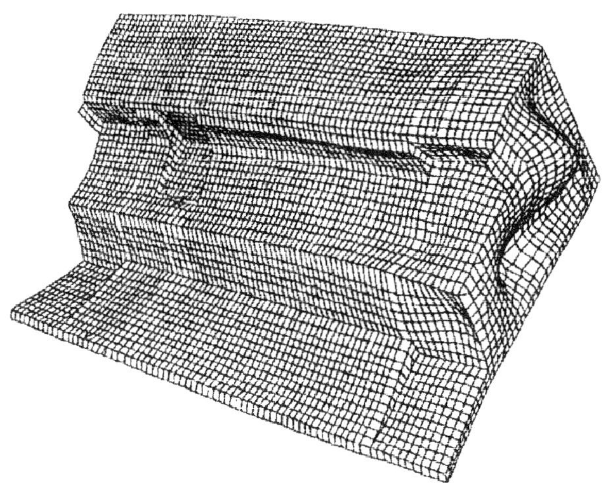

FIGURE 1. CORNER DROP TEST OF LAPTOP COMPUTER

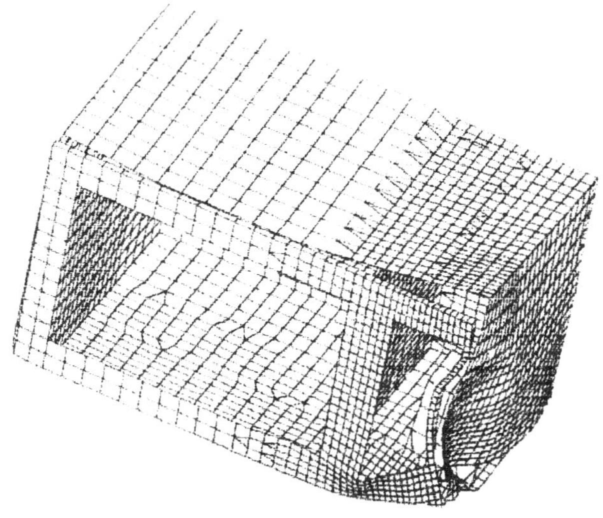

FIGURE 2. CORNER DROP TEST OF UNPROTECTED MICROWAVE OVEN

and mechanical components can survive the shock event.

Designing consumer appliances and packing to survive shocks raises a challenge to cover the large variety of shock configurations. The heights of free fall can be different. The ground surfaces may have different hardness levels with subsequent different behaviors. Also the attitudes at impact need be multiplied especially when one or several protruding parts are included in the design of the appliance. Different shock configurations may appear to be critical for structural integrity and electronics survival. Finally, if multiple shocks is an applicable survival criterion, the sequential order of the shock configurations may need to be explored as well.

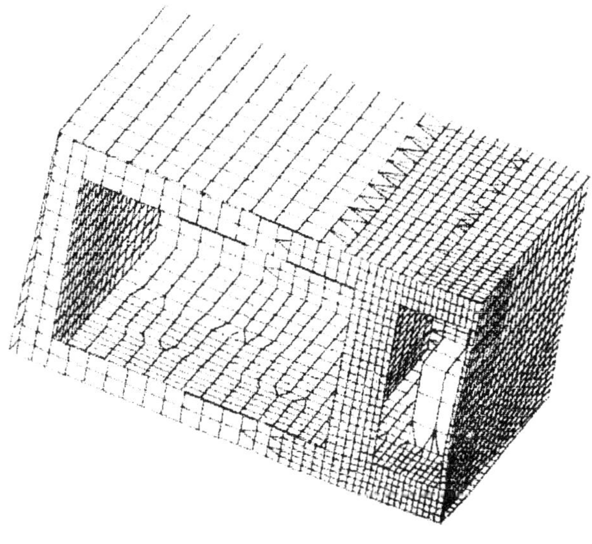

FIGURE 3. CORNER DROP TEST OF PROTECTED MICROWAVE OVEN

In the face of that challenge, an ad-hoc CAE application allows to carry out shock simulations very early in the design cycle, thus eliminating costly iterations between design and prototype testing. It also allows to carry out a great number of shock simulations for many different configurations. And last, it would provide the designer very detailed results on the behavior of each component during the shock. This would enable the designer to understand the reaction to shock and to optimize the design of the appliances.

2. Explicit Finite Element

The explicit finite element method (Haug et. al., 1989) combined with the industrial experience and technical expertise paved the way for the integration of numerical tools in structural crash simulation up to the design process level. This integration process has started in the mid 80s in the automotive industry and by the early 90s no significant automaker was designing cars without using numerical simulation as a design tool to optimize the structure and comply with the crashworthiness regulations. A similar trend has followed among the rolling stock manufacturers, the shipbuilders and other vehicle manufacturers.

The growing demand for similar analyses in the consumer appliance industry is guiding the adaptation and effective usage of specialized tools for shock hardening analyses for consumer appliances. Examples of such applications are depicted on Figures 1 to 4. The corner drop of a laptop computer (Kisielewicz et. al., 1993) is illustrated on Figure 1 with deformation amplified by a factor of 10 for a better readability. Figures 2 and 3 show the deformation of a microwave oven (Kisielewicz et. al., 1992) following a corner drop at 5 m/s without and with pads of EPS (the latter are not depicted on the figure) clearly illustrating the

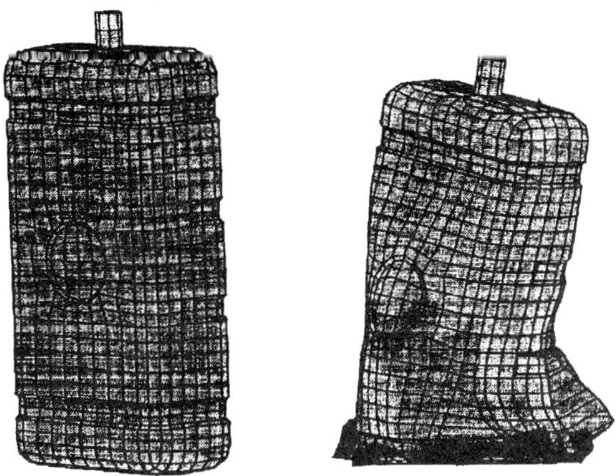

FIGURE 4. DROP TEST OF LIQUID FILLED PLASTIC RESERVOIR

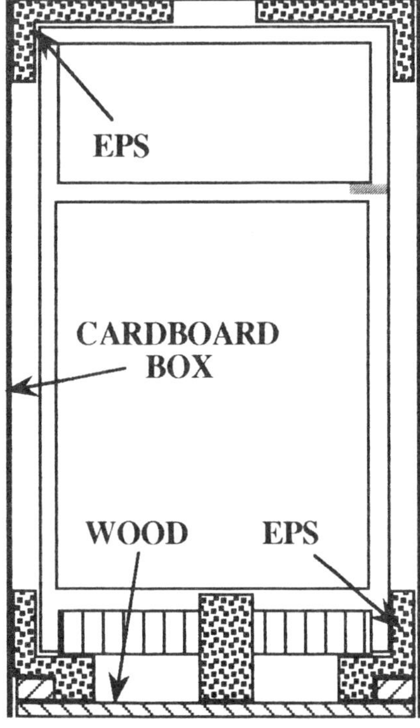

FIGURE 5. PACKING OF THE REFRIGERATOR

dramatic protection provided by such energy absorbing pads. Other examples of applications include cameras, nuclear spent fuel casks (Diersch et. al., 1993), or liquid filled plastic reservoir

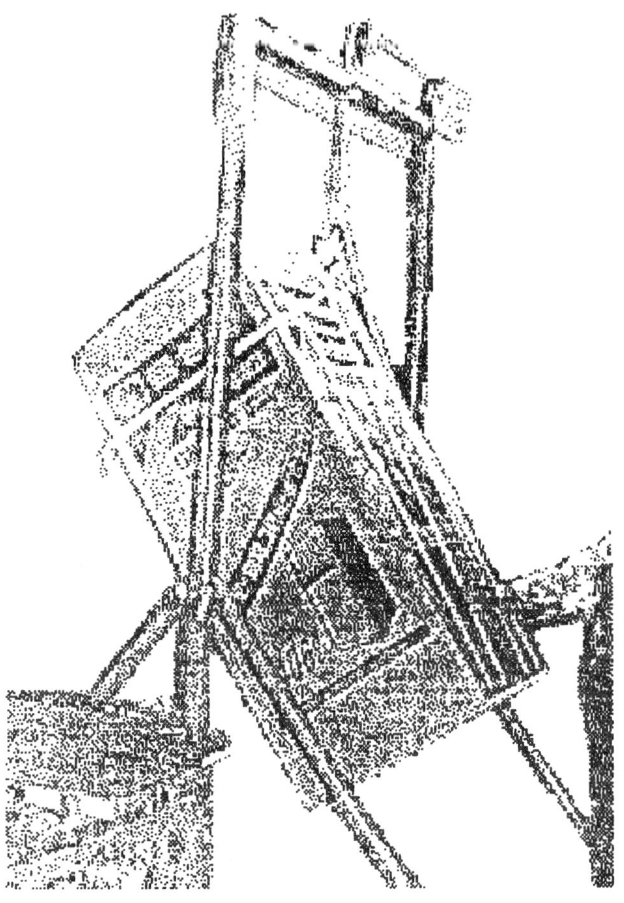

FIGURE 6. DROP TEST SETUP

shown in Figure 4.

APPLICATION CASE

1. Refrigerator case

Goldstar routinely carries out drop tests of packed refrigerators in 8 configurations (4 corner drops and 4 side drops). The project reported here is aimed at demonstrating the feasibility of using numerical simulation to predict the behavior of the refrigerator and its packing during such drop tests. The configuration selected is a lateral side drop.

The simulated drop test of the packed 406-liter refrigerator was from a height of 50 cm (impact velocity of 3.13 m/s) on its lateral side. The principle of packing is depicted on Figure 5. The physical drop test setup is illustrated on Figure 6 and a typical damage following such drop test, i.e. an undesirable indentation, is illustrated on Figure 7.

2. Modeling

The packed refrigerator was modeled using about 10,000 nodes,

FIGURE 7. DAMAGED REFRIGERATOR AFTER DROP TEST

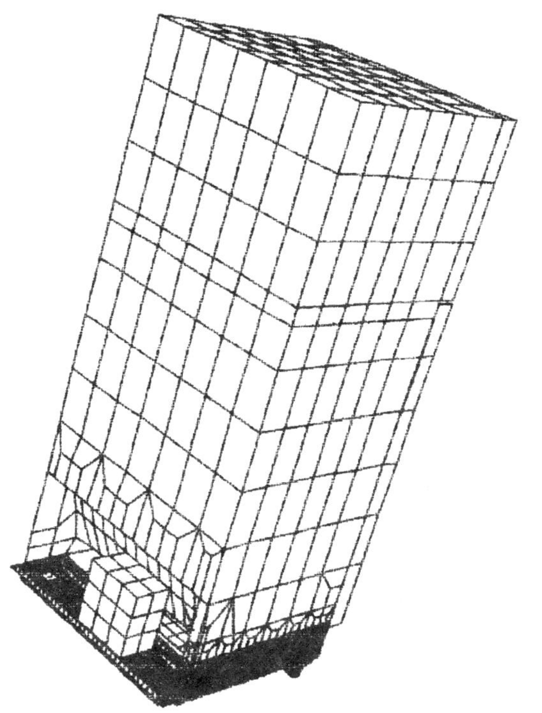

FIGURE 8. MESH OF REFRIGERATOR

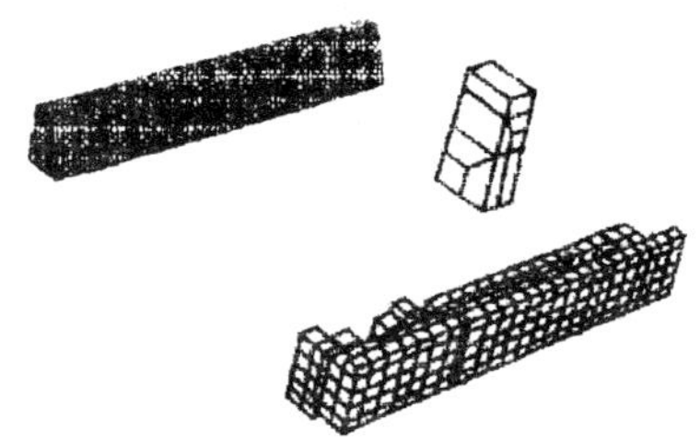

FIGURE 9. MESH OF EPS PADS

FIGURE 10. MESH OF WOODEN SUPPORT

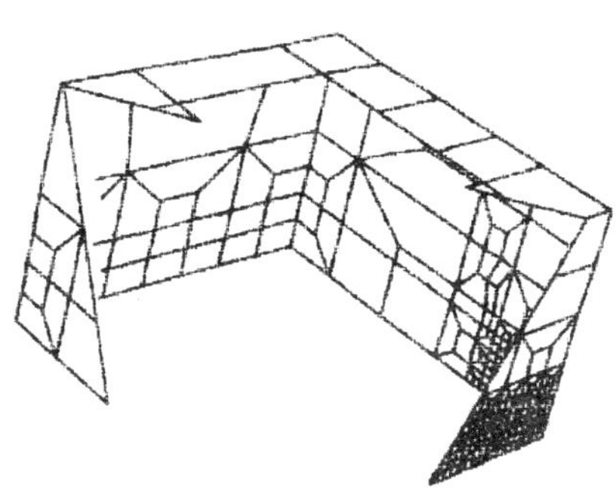

FIGURE 11. MESH OF CARDBOARD (BOTTOM PART)

5,000 solid elements and 5,000 shell elements. Figures 8 through 11 depict the mesh of several parts of the packed refrigerator.

The refrigerator was modeled with its main structural parts especially in the bottom part but also in the upper part when it was deemed that these members provided some stiffening to the main body like the shelf between the ice compartment and the cooling room. The door was disconnected from the main body except at the locations of the hinges.

The equipment contained in the refrigerator shown in Figure 12 included a heavy compressor, a transformer and a number of tubing and wiring. The compressor and transformer were modeled rigidly and connected to their support plate. The wiring and copper tubing are very flexible (expansion loops) and were not included in the model.

The EPS pads were modeled using different schemes for the pad in the area of the impact line (more detailed) and the pads away from it and the one supporting the door (coarser mesh). The wooden support was modeled using three rows of solid elements in the area close to the impact point in order to capture crushing and bending effects during the drop test. The cardboard box was modeled in detail in its bottom part.

FIGURE 12. REFRIGERATOR EQUIPMENTS

Load

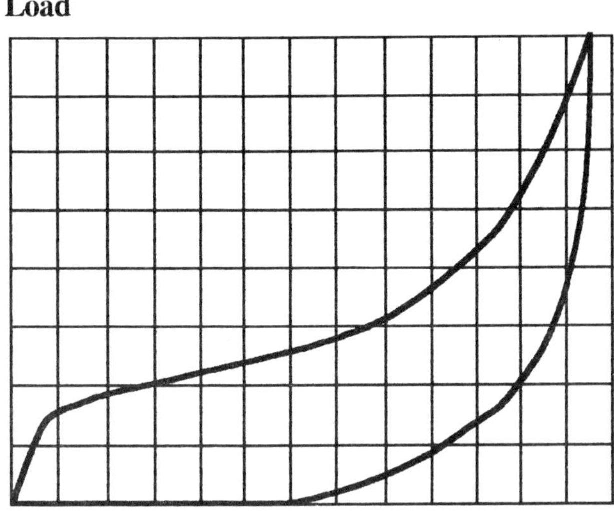

Deformation

FIGURE 13. TYPICAL LOAD-DEFORMATION CURVE OF EPS

Material calibrations were carried out for all non standard materials, i.e. wood, EPS, polyurethane foam, nylon straps, and cardboard. Figures 13 and 14 depict typical test results on samples of EPS and polyurethane foam used to calibrate the material models (crushable foam material type). Wood was modeled as elastic-plastic as during the drop test no rupture was experienced (very limited energy is absorbed in the wooden support). Cardboard was modeled using elastic-plastic shell element with approximate properties. Further studies for a more refined calibration were deemed useful but postponed to a further stage of the project.

3. PAM_CRASH™
The simulation of the drop test was carried out using PAM_CRASH™. PAM_CRASH™ is a 3D Lagragian code using an explicit finite element scheme for analyzing non linear dynamic structural responses. It is widely used in the automotive industry as well as by other vehicle manufacturers and a number

Load

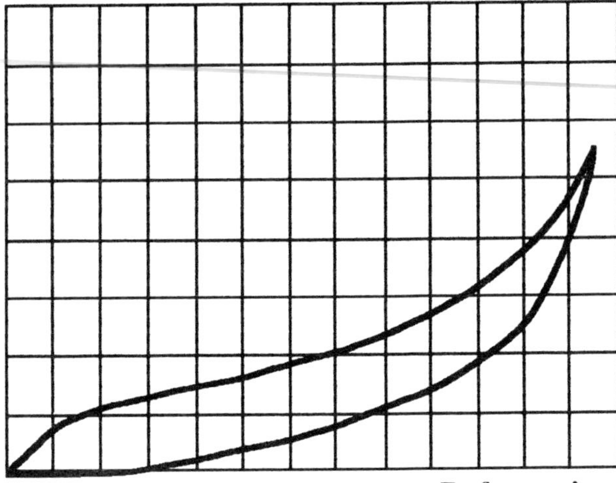

Deformation

FIGURE 14. TYPICAL LOAD-DEFORMATION CURVE OF POLYURETHANE

of appliance manufacturers.

Spatial discretization can be achieved using solid, shell or beam elements. A variety of material models are available to model linear, nonlinear, isotropic or anisotropic, strain rate dependent and material behaviors including their failure. Contact algorithms allow to represent gaps and sliding contacts between parts of a structure.

PAM_CRASH™ has been fully validated using experimental results for a wide variety of cases.

4. Results
The final deformed shape of the refrigerator and EPS is depicted on Figure 15. Some zooms are provided on Figures 16 through 18.

Absorbed energies are illustrated on Figures 19 and 20. One can see on these time histories the large difference (about 1.5 order of magnitude) between the levels of energy absorbed in the EPS pads (designed for such purpose) and the other parts of the refrigerator and other packing parts. While the amount of energy in the latter is limited it is nevertheless sufficient to produce the undesirable effects on the refrigerator. These limited amount of energy are concentrated in limited areas where they correspond to irreversible deformations such as the indentation produced during the physical drop tests as shown in Figure 21.

From Figure 20, one can follow the major steps of the deformation of the packed refrigerator. The first part to react with its rather high stiffness and large elastic reservoir are the PP bands. These load all the other parts by transferring the accumulated energy and deforming them in an irreversible way. Some elastic energy is also stored in other parts of the refrigerator and transformed into kinematic energy (rebound like movement) of the appliance.

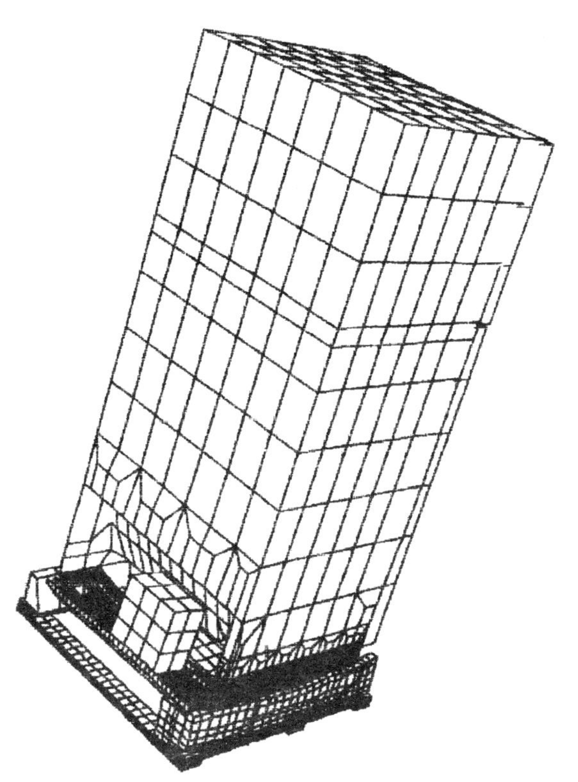

FIGURE 15. GLOBAL VIEW OF FINAL DEFORMED SHAPE

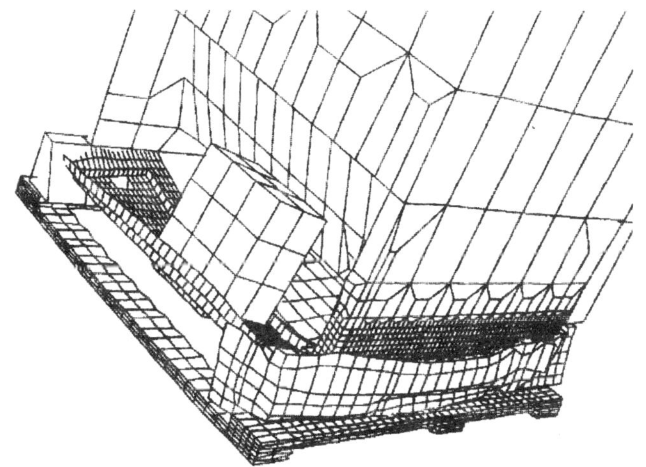

FIGURE 16. LOCAL VIEW OF FINAL DEFORMED SHAPE

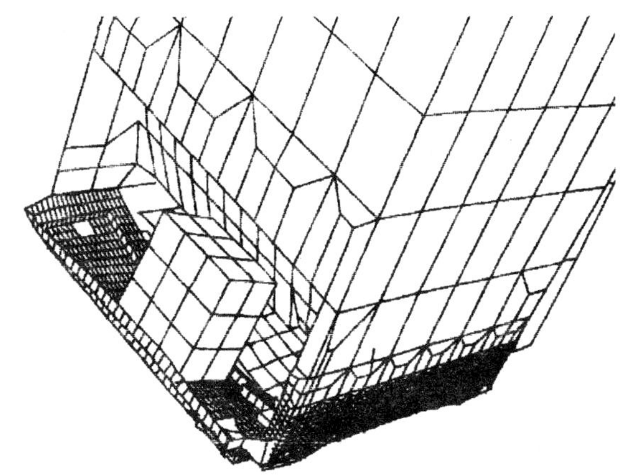

FIGURE 17. FINAL DEFORMED SHAPE OF BASE OF REFRIGERATOR

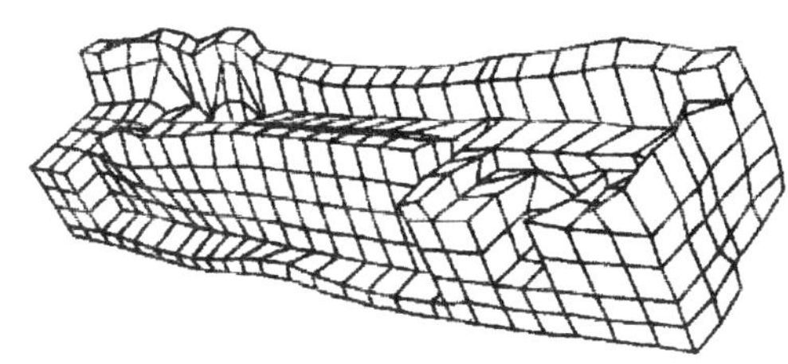

FIGURE 18. FINAL DEFORMED SHAPE OF FRONT OF EPS PAD

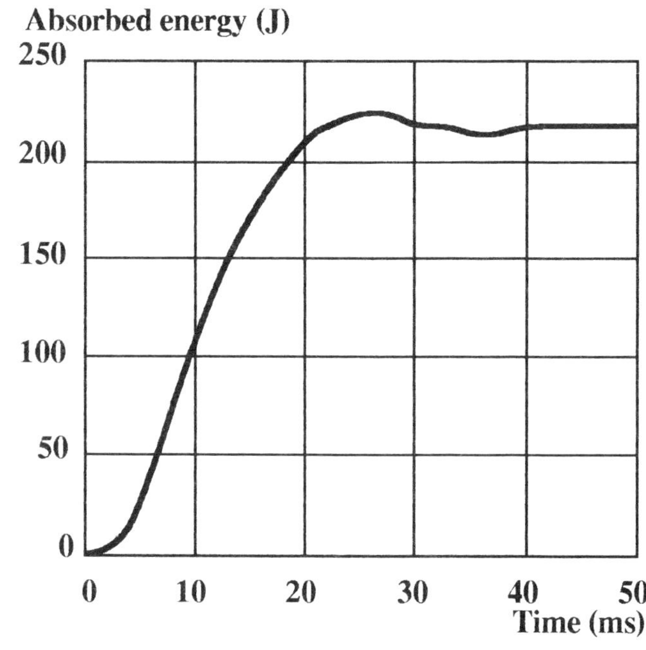

FIGURE 19. INTERNAL ENERGY IN EPS PADS

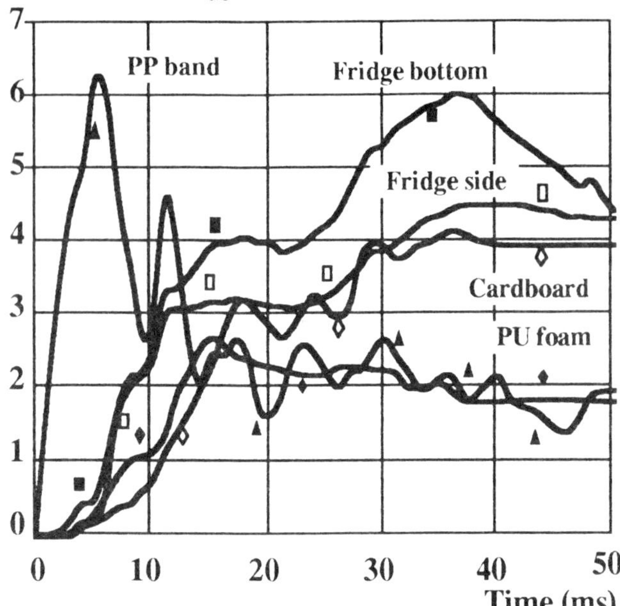

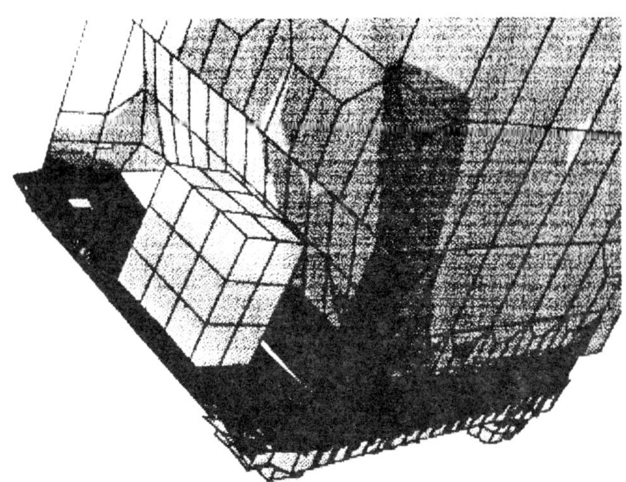

Absorbed energy (J)

FIGURE 20. INTERNAL ENERGIES IN SEVERAL PARTS OF REFRIGERATOR AND PACKING

FIGURE 21. VON MISES STRESS DISTRIBUTION ON REFRIGERATOR

RECOMMENDATIONS

The key issue in obtaining accurate and reliable results by using numerical simulation for predicting structural behavior during drop tests is the calibration of material. This is especially true because most materials used in manufacturing appliances and used in packing them are not analyzed in a standard way as ductile materials (steel) are. Proper material model and calibration procedure are needed in order to obtain the required level of accuracy.

Crushable foams (EPS, polyurethane, etc.) are very common in packing and even though they need to be replaced by new packing solution, it is necessary to establish reference position in order to study these new packing solutions.

Other common materials used in packing are wood and cardboard. While some results are available regarding their non-linear dynamic behavior, proper material model are still lacking in simulation packages as well as accepted material properties.

It is thus recommended that a data bank of material properties for these packing materials be developed. Such data bank must be linked to some specific material models, some of which are still lacking.

CONCLUSIONS

An engineering project aiming to demonstration of the feasibility of using numerical simulation for predicting structural behavior during drop tests of packed appliances has been presented. This field of application of numerical simulation techniques widely used in designing vehicles is now emerging very quickly under the pressure of coming regulations related to the packing of artifacts.

The approach followed in modeling a packed refrigerator for simulating a lateral side drop has been explained and the main results obtained from the simulation have been described.

Some recommendations about the needs to be addressed for an effective usage of numerical simulation in that field have been drafted and should lead to further projects in that field.

REFERENCES

Kadoya, T., 1990, "Food Packaging," Academic Press, Harcourt Brace Jovanovich Publishers, San Diego.

Haug, E., Clinckemaillie, J., Aberlenc, F., 1989, "Computational Mechanics in Crashworthiness Analysis," Post-Symposium Short Course of the 2nd International Symposium on Plasticity, Nagoya, Japan.

Kisielewicz, L. T., Ando, K., Endo, M., Mita, K., Trameçon, A., 1993, "Numerical Simulation of Laptop Computer Drop Tests," PUCA '93 PAM Users Conference in Asia, Shinyokohama, Japan.

Kisielewicz, L. T., Petitjean, A., 1992, "Numerical Simulation of Drop Test on Microwave Oven," ESI Asia internal paper.

Diersch, R., Spiker, H., Hüggenberg, R., Rückert, J., 1993, "Strength Behavior of Transport and Storage Casks for Spent Nuclear Fuel under Drop Test Conditions," PAM'93 3rd European Workshop on Advanced Finite Element Simulation Techniques, Schlangenbad, Germany.

PAM_CRASH™ User's Manual, PSI, Paris, France.

AUTHOR INDEX

**DE-Vol. 87, Reliability, Stress Analysis, and Failure Prevention
Issues in Emerging Technologies and Materials**

Aoki, Mitsuhiro 35
Baylor, Jeffrey S. 41
Cho, Wonjoon 109
Guiqin, Wang 95
Gür, Mustafa 49
Huang, Yuan Mao 79
Huffaker, A. Vincent 67
Jian, Ma...........95
Katsuo, Masahide 19
Kawawaki, Masataka 19
Khutoryansky, Naum 11
Kim, Seoggwan 109
Kisielewicz, L. T. 109
Maguire, John F. 57, 63
Maimon, Oded Z. 67
Matsumura, K. 109
Miller, Michael A. 63
Molin, Wang 89
Nakano, Yuichi 19, 25
Nishikawa, Osama 35
Petitjean, A. 109
Sancaktar, Erol 1, 41, 49
Sawa, Toshiyuki 19, 25, 35
Sosa, Horacio 11
Talley, Peggy L. 57
Temma, Katsuhiro 25
Turgut, Aydin 49
Uchida, Hiroaki 25
Wei, Yong 1
Xia, Houchun...........99
Xue, David Y. 99
Yazhou, Jia 89, 95
Zhixin, Jia 89, 95

Book Number: H01035